THE CENTRE FOR FORTEAN ZOOLOGY
expedition report 2007
guyana

Typeset by Jonathan Downes, Edited by Jon and Corinna Downes
Cover and Layout by Hennis y *El gato orango* for CFZ Communications
Using Microsoft Word 2000, Microsoft , Publisher 2000, Adobe Photoshop CS.

Photographs © 2008 CFZ except where noted

First published in Great Britain by CFZ Press

**CFZ Press
Myrtle Cottage
Woolsery
Bideford
North Devon
EX39 5QR**

© CFZ MMVIII

All rights reserved. Without limiting the rights under copyright reserved above, no part of this publication may be reproduced, stored in or introduced into a retrieval system, or transmitted, in any form of by any means (electronic, mechanical, photocopying, recording or otherwise), without the prior written permission of both the copyright owners and the publishers of this book.

ISBN: 978-1-905723-25-6

CONTENTS

- 7. *Introduction*, by Jonathan Downes
- 11. *Foreword* by Dr Karl P.N. Shuker
- 17. Richard Freeman
- 41. Paul Rose
- 57. Dr Chris Clark
- 65. Jon Hare
- 71. Lisa Dowley
- 111. Photo section
- 157. *Giant snakes in South America* by Richard Freeman
- 175. *Keeping the home fires burning* by Jonathan Downes
- 207. *Doing it for the kids* by Jonathan Downes
- 221. Press releases
- 231. A selection of press cuttings
- 235. Capcom - our sponsors

INTRODUCTION

By Jonathan Downes
(Director, Centre for Fortean Zoology)

The Centre for Fortean Zoology is the only professional, scientific and full-time organisation in the world dedicated to Fortean Zoology; a portmanteau discipline which includes cryptozoology - the study of unknown animals. Since 1992, when it was founded, the CFZ has carried out an unparalleled programme of research and investigation all over the world. Because we are funded purely from private subscription, we feel that it is only appropriate that we make all our findings and research public as well. This is the second, of what we hope will be a long line, of CFZ expedition reports published in book form, and as Director of the CFZ it gives me great pleasure to write this introduction.

This expedition was a groundbreaking one for us on a number of levels.

Firstly, it was our first foray to the continent of South America - a place which the early explorers described as the `Green Hell`. Our destination was the country of Guyana, and in many ways I was sad not to be going there because it is a country that I have always wanted to visit. My childhood hero Gerald Durrell wrote infectiously about some of his animal collecting trips there in the late 1940s and early 1950s when it was still the British Crown Colony of British Guiana, and these books instilled me with a determination to one day visit there.

Well, I might not have made it in person, but being on the other end of an increasingly dodgy SatPhone link, was the next best thing.

Secondly, it was the first expedition for which we had managed to attract commercial sponsorship. Our sponsors were CAPCOM - one of the world's leading video game manufacturers, and they wanted to publicise their game `Monster Hunter Freedom`. They had shoved the words `Monster Hunter` into Google, and my name had come up, mainly because I had written my autobiography with this very title some years before.

Would we be interested in a sponsorship deal? They asked diffidently. Of course we would! Anyway, by all accounts, the Monster Hunter games are great fun, and a worthy piece of product for us to promote, so the game was afoot.

Thirdly, on many levels this has been our most successful to date, because, although unfortunately we were unable to carry out one of the mission's main objectives; to search for giant anacondas, had to be aborted, we came back with a wealth of information on cryptids about which we had never even heard before. To the best of our knowledge, this book is the first time that anything has been published on the subject of the tiny red faced pygmies of Guyana, or the strange dwarf red caiman. So, on this level, the expedition was a resounding success.

With this volume we have stuck with the format that debuted with the Gambia expedition report in 2006; publishing accounts of the expedition from each of the team members. I wrote in that report:

"We decided that rather than editing all the different accounts into a single narrative, it was better to provide all five accounts separately. This we have done, and whilst every effort has been made to standardise all spellings and grammar, it is almost certain that some anomalies will remain. However, it makes for interesting reading. Just as in *Rashomon* - a 1950 Japanese motion picture directed by Akira Kurosawa (in collaboration with Kazuo Miyagawa) and starring Toshiro Mifune.- the same story being told from a variety of different perspectives, allows us a uniquely fortean view of events.

Some accounts are matter-of-fact and scientific, others more subjective, but they add up to a fascinating whole. "

For those of you not aware of *Rashomon* it is a play and a film based on two stories by Ryūnosuke Akutagawa (1892-1927), it describes a rape and murder through the widely differing accounts of four witnesses, including the perpetrator and, through a medium, the murder victim. The story unfolds in flashback as the four characters - the bandit Tajōmaru (Mifune), the murdered samurai Kanazawa-no-Takehiro (Masayuki Mori), his wife Masago (Machiko Kyō), and the nameless Woodcutter (Takashi Shimura) - recount the events of one afternoon in a grove. But it is also a flashback within a flashback, where the woodcutter or priest has told what each individual said at the court. Each story is mutually contradictory, leaving the viewer unable to determine the truth of the events, and should be required viewing for every fortean.

In this volume, our multi-faceted approach - I feel - works better than ever. Richard Freeman looks at the events from the perspective of a cryptozoologist, Chris Clark from the perspective of an explorer. Lisa Dowley is awed by the experience, Paul Rose - a well known humorous journalist in his own right - brings his own unique worldview into play, and the punishing heat of what Richard Freeman describes as Guyana: The Savage Land apparently turned Jon Hare into William Burroughs!

Five totally unique explorers present five totally unique viewpoints on their experiences.

In addition we have included an excerpt from Richard Freeman's classic book *Dragons: More than a Myth?* (CFZ, 2005) on giant snakes in South America, a few notes from yours truly on the experience of manning the expedition base in North Devon, some notes on the didi and water tiger, and a foreword by our friend and colleague Dr. Karl Shuker.

I hope that you agree with me that it was all well worth it!

Jon Downes,
Director, CFZ
Woolfardisworthy,
North Devon
February 2008.

FOREWORD

By Dr Karl P.N. Shuker

"Now, down here in the Matto Grosso" – he swept his cigar over a part of the map – "or up in this corner where three countries meet, nothin' would surprise…"

Sir Arthur Conan Doyle – The Lost World

The above lines are quoted from the world's most popular cryptozoological novel, but are very relevant to present-day reality too. Since the early 1990s, an extraordinary array of significant new creatures has been discovered in South America. These include a seemingly never-ending series of new monkeys, the world's largest species of peccary, a new coati, a dwarf manatee, an extra-large paca, and even a distinctive tapir – hardly the smallest or least inconspicuous of beasts, one would have thought.

In addition, during this same period a few bold, open-minded zoologists have bravely run the gauntlet of potential criticism from cryptozoological sceptics, and have sought even more exotic mystery beasts amid the immense verdant wildernesses still awaiting detailed scientific exploration within this vast tropical continent. Dr Peter Hocking, for instance, has scoured the jungles of Peru in search of several different types of scientifically-unclassified felid as well as a chimpanzee-sized crypto-primate known locally as the isnachi. Similarly, Dr David Oren has made a number of forays into the heartlands of Brazil in search of the fearsome mapinguary, which he believes may well be a surviving species of ground sloth. Also in Brazil, Dr Marc van Roosmalen has uncovered an astonishing diversity of seemingly new mammalian forms, including many of those Neotropical examples officially recognised by science during the past few years.

Now, a major new cryptozoological expedition, sent by the indefatigable CFZ, has succeeded in amassing a very extensive archive of very exciting, brand-new data from a very different region of South America – the former British colony of Guyana, which until now had attracted little cryptozoological attention in comparison to this continent's larger and more famous countries. Yet as you will discover when reading this fascinating book, the team has gathered tantalising evidence for the existence here of several very noteworthy creatures presently undescribed by science. These include an incredible aquatic beast known locally as a water tiger, an elusive red-faced humanoid entity, a truly gargantuan form of anaconda, the ape-like didi, and a minuscule crocodilian.

I shall not spoil the fun by giving any further details here, but rest assured that you will not be disappointed by the revelations in store as you read this unique book – which records in detail, by the team members involved, what is certainly some the most significant cryptozoological field work to have been conducted anywhere in South America for many years.

Readers of Heuvelmans's books, my own, and those of other cryptozoological chroniclers will be well aware that a stunning selection of mystery beasts have been reported from this continent down through the centuries - with such thrilling creatures as the gigantic worm-like minhocão, the pack-hunting Warracaba tiger, various long-necked dinosaurian cryptids, a reclusive mystery bear called the milne, the infamous Patagonian plesiosaur, the vertically-undulating huillia of Trinidad, the giant toothless lake shark (a catfish?) of Paraguay and Bolivia's feline dog (or canine cat?) known as the mitla both reported by Fawcett, the voracious iemisch, and the shy striped mystery cat with extra-large fangs from Colombia among the more memorable.

Unfortunately, all of these cryptids are linked by the sad fact that no-one has seriously looked for them in a long while. And who knows? Even if they did once exist, with the relentless destruction of South America's rainforests they may by now have become extinct – lost to science before it had ever recognised their existence. This is why modern-day cryptozoological expeditions such as the fine example documented here are so unusual, but so important. Unusual inasmuch as they have even took place at all, let alone have proven so successful in obtaining new data, and important because they are highlighting the very real possibility that in this vanishing rainforest domain, there are indeed dramatically new creatures still awaiting discovery and, with that discovery, the protection that one sincerely hopes it would bestow upon them.

So read this book with delight and enthusiasm, as I have done, and let us commend the CFZ for a job very well done - confirming, as I believe a certain French Father of Cryptozoology once wrote, the great days of zoology are not done.

Karl Shuker,
West Midlands,
24 February 2008

Map of Guyana

Sint Maarten-Barbados-Guyana (Jan. 2005)

F = Fähre
S = Frachter

VENEZUELA

ATLANTIC OCEAN

- Mabaruma
- Shell Beach
- Amakuru River
- Kaituma River
- Barima River
- Waini River
- Kwebanna
- Pomeroon River
- **CHARITY** S
- Anna Regina
- Lake Mainstay
- Charaishiru
- **FORT LAND**
- **GEORGETOWN**
- Emerald Tower
- Timberhead
- Arrowpoint
- Cuyuni River
- Kartuni River
- Parika
- Ogle
- Cheddi Jagan International Airport 6 Std.
- 42
- Sakaika Falls
- Puruni River
- Kyk-Over-Al
- **BARTICA**
- 104
- **NEW AMSTERDAM**
- Essequibo River
- Mazaruni River
- Baracara
- Berbice
- Mahaicony River
- Abary River
- Nieuw Nickerie
- Ekereku River
- Kamarang River
- Kaburi River
- Demerara River
- **LINDEN**
- Berbice River
- Canje River
- **CORRIVERTON**
- 113
- **Paramaribo Suriname** (March 2003)
- **Mt Roraima**
- 109 Ituni
- **Pakaraima Mountains**
- Potaro River
- **SURINAME**
- Kaieteur Falls
- Ikanami Falls
- Mabura Hill
- Mandia
- Canister Falls
- 100
- Orinduik Falls
- Burro Burro River
- **F**
- Kurupukari
- Govenor Falls
- Govenor Lights Falls
- **Iwokrama International Centre**
- Wonosoro Falls
- Rockview Lodge
- **ANNAI**
- Tiger Falls
- **Kanuku Mountains**
- Karanambu
- Karanambu Ranch
- 112
- Rupununi River
- King Frederick William IV Falls
- King William IV Falls
- Barrington Brown Falls
- King George VI Falls
- King Edward VII Falls
- Flows south to Amazon
- **LETHEM**
- Rewa River
- Rupununi Savannahs
- Essequibo
- Oronoque Falls
- Dadanawa
- Dadanawa Ranch
- Great Falls
- Takutu
- Kuyuwini River
- New River
- Sir Walter R Falls
- **Venezuela** (Sept. 2002)
- **Marudi Mountains**
- Oronoque River
- Boa Vista
- Kassikaityu River
- Wabu Falls
- **BRAZIL**

KEY

- Waterfalls
- Nature Resort
- Major Town/Settlement
- Historic Site
- Precious Mining
- Airstrip
- Amerindian site

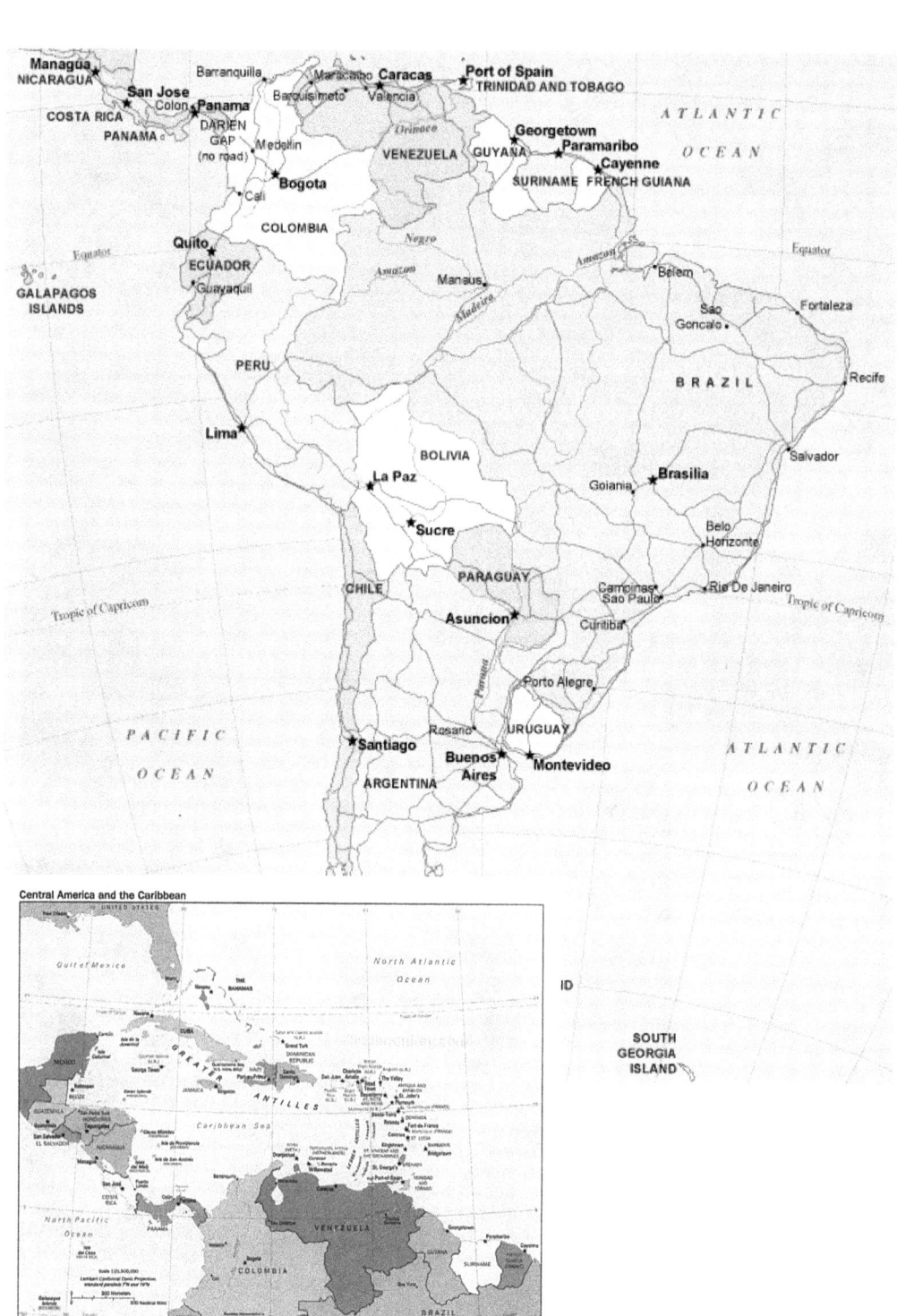

RICHARD FREEMAN

The Centre for Fortean Zoology's expedition to Guyana had its genesis with an entry in Michael Newton's excellent '*Encyclopaedia of Cryptozoology*' on crypto-tourism. In this book, he mentioned a company called 'Guided Cultural Tours', which offered expeditions in search of giant anaconda in Guyana.

A man called Damon Corrie runs 'Guided Cultural Tours', and he told me that the company specialises in showing people the true Guyana of the native peoples. Damon is a chief of the Eagle Clan Arawak Amerindians, and as well as being a well respected figure to the native peoples, he also is a conservationist who breeds and studies Guyanese reptiles and invertebrates.

Damon told me that only last year a gigantic anaconda had been seen at a remote pool, known as Corona Falls. He had spoken with the hunters who had seen the beast, and they had told him that it was so large that they had fled from it. When he asked how big the snake was one, of the men pointed to a 30 foot palm tree. He told Damon that a dead tree of the same size had been lying in the water. The anaconda was crawling over it and its head and tail extended beyond the ends of the tree. This would make the snake around 40 feet long.

Damon also mentioned the di-di, which is a large hairy creature seemingly akin to the yeti or sasquatch that had been seen in Guyana, and also told me more vague stories of dragons that are said to inhabit the mountains. The CFZ decided to mount an expedition in search of these creatures.

For the first time ever, we were able to secure some funding. Sam Brace, who worked for the computer games company Capcom, tied in our expedition to the release of his company's game 'Monster Hunter Freedom'. Capcom kindly donated £8000 towards the project. This was a major breakthrough for the CFZ. Time and again we have been let down, in particular by television companies and stations. We get around twenty TV companies approaching us every year with a view to making a documen-

tary, or even a series about our work. Every time it ends up the same way. The company comes to meet us, and we discuss projects. They go away, excited, back to the station executives who flatly refuse to make the shows. In the past we have been refused for being 'too real' i.e not enough to do with flying saucers, healing crystals or rescue mediums. Another time we were told that the company would not back us unless we could guarantee we would find the creature we were looking for! Yet another time a person from the BBC expected us to finance an expedition that they would send a camera crew on, and get a film out of. They said they would give us £50. We told them exactly where to go.

I have a long history of being jerked about by TV companies. Back in my student days, I was trying to get an expedition together to search for the thylacine (*Thylacinus cynocephalus*) or marsupial wolf, in Tasmania. I was contacted by a company that had done several documentaries. They had heard of my plans and wanted to make a film about the search for the thylacine, so I twice travelled down from Leeds to London for meetings at their offices. They wanted to include the expedition in the series '*To The Ends of The Earth*' series. I told them all I knew about the thylacine - where to look, what to do and what not to do. They repaid this by sending someone else to Tasmania on the back of my knowledge. Their excuse? The producer of the series wanted to use a pretty girl, as all the other expeditions were being led by me. They sent a vacuous child instead of an experienced zoologist just because of looks. Needless to say the expedition never got near the creature. On another occasion, an American TV company executive wanted me to take scantily clad women in low cut tops on the expedition; as '*sex sells*'.

I'm very bitter about the way TV has treated us in the light of the excrement that companies spew out as 'entertainment'. Wall-to-wall reality TV, property shows, house garden makeovers, dozens of antique shows and countless cookery programmes. The few TV gems left are lost in a sea of odure and banality.

It came as quite a surprise when Capcom approached us with this idea. I had never even considered approaching a games company for sponsorship. The whole thing came from left field. The idea was the brainchild of Sam Brace, who has now sadly left the company. I can only hope that there are other open-minded people like Sam out there, who truly think outside the box.

We were approached in the weeks before we left by two TV production companies telling us they were interested in making a film about our expedition. As always nothing came of it.

As well as myself, the team included Dr Chris Clark and Jon Hare who had been on several past expeditions, Lisa Dowley, who had joined us on the previous year's trip and was acting as photographer and archaeologist, and Paul Rose, aka Mr Biffo. It was Paul's first such excursion. He is a TV writer, author and most importantly the man behind the sadly defunct 'Digitiser' on teletext. It was a collection of bizarre non-sequiturs that had Jon Downes and I in tears of mirth for several years.

To give you an idea of Mr Biffo's sense of humour here are some of his more comical jokes.
Q. What do you call a giant killer bat?
A. Super Beast 47.
Q. What do you call a man who thinks that dogs are pigs, pigs are icebergs, and icebergs are ice cream?
A. Helios 7
Q. What do you get if you cross ice lollies with onions?
A.   Iced punions.

Luckily on our outward flight, we had quite plush seats with televisions placed in the back of the seat in front. I happily passed the long flight to New York watching episodes of *Dr Who*, *The Mighty Boosh* and Michael Moore's excellent film *Sicko*.

In the paranoid and unwelcoming climate of US officialdom post-9/11, we had to fill out paperwork even though we were not even staying in the country. I was glad to get out on the next leg of the flight.

Upon our arrival in Guyana's Georgetown airport we were met by Damon, who took us to his native village of Pakuri. I was glad to get out of Georgetown. On the way into Guyana, I only saw a little of it, but it looked filthy, backward and mean. I have been in some nasty, dirty places, but Georgetown had a vile *genius loci*. Not only was it ugly, but it seemed dangerous. There was an air of pent up aggression about the place that seemed ready to explode at any moment. We didn't see any violence, but it felt as if a riot was simmering just beneath the surface and would explode for no reason at any time. It was a long and bumpy ride in an open-backed truck to the village. Pakuri is also known as St Cuthbert's Mission, but Damon, who is a champion of Amerindian Rights, encourages all to call the village by its original name. It takes its name from the pakuri tree. Damon showed us the last such tree in the area, the others having long since been cut down. I made a point of always calling the village `Pakuri`, as I hate missionaries destroying native culture with their pernicious twaddle. The Amerindians have already suffered at the hands of the Europeans back in Columbian times. Then black immigrants who were brought over first as slaves but together with Asians became the main population of the cities such as Georgetown marginalised them further. The blacks now run the country and the government, with the indigenous peoples having almost no rights. Now it seemed that their much-reduced population was being yet more marginalised. Having western fairy tales destroying their religion was the last straw!

Pakuri seemed like a content and stable community, unlike the filthy and crime-ridden Georgetown. We took a swim in a nearby creek, the waters of which is stained red by the tannins from the leaves of the plants along its banks. Once in the wine coloured waters it gives the illusion of turning your skin red. Damon told us that small caiman and anaconda were sometimes seen in the creek, but, thankfully, the infamous candiru (*Vandellia beccarii*) was absent from the waters. The candiru is a tiny parasitic catfish. It evolved to swim into the gills of other, larger fish, attach itself via spines on its head, and proceed to eat the victim from the inside out. It is notorious for what it does to human beings. The candiru is excited into a frenzy by any traces of urine in the water. If a bather is naked it will attack the genitals, wriggling into sexual openings, both male and female, and lodging itself there with its backward pointing barbs. It cannot be pulled out of its flesh den without extreme pain to the victim, as well as serious damage. Once inside a penis or vagina it will start merrily munching away like a piscine Hansel or Gretel in a cottage made of meat rather than of gingerbread. Frankly I'd rather fall victim to an anaconda. At least that is a predator with a sense of decorum and less eye-watering habits.

Whilst in Pakuri, we were told of a di-di encounter that had occurred only two years before. It happened in another Amarindian village some 30 miles north of Pakuri, when two children - a boy, and a girl of about 12 - were walking home from school across the savannah. What the boy described as a 'huge hairy man' stepped out of a stand of trees, and grabbed the girl. She was never seen again. There was no police investigation, but this is not surprising as the government of Guyana seems to care very little for its native peoples.

However, the story of the hairy giant kidnapping a woman was one we would hear again whilst in Guyana and one that is told around the world where-ever these hominids are seen.

We heard about another man from Pakuri who, several years previously, had seen a di-di walking away from him, but unfortunately this man was not in the village at the time, so we could not interview him.

Damon showed us a rainbow boa *(Epicrates cenchria)* that he had captured, which was of a rusty colour phase that I had not seen before. We were also shown a new species of green scorpion discovered by Damon, which was so new that it had not been officially described, nor had it been given a scientific name.

Damon's brother-in-law, Foster, told us that several years ago, in a watercourse a few miles from the village, he had seen the trail of a big anaconda. Judging by its width, the creature that had made it would have been far larger than the 20 foot stuffed specimen in the National Museum in Georgetown.

We travelled back to the unpleasant environs of Georgetown to catch a bus inland to Letham, where the expedition proper would begin. We spent a number of dull hours sitting around a depressing and grubby station in the ugly city waiting for the bus. When it arrived our hearts fell. Looking as if it was held together by rust, Sellotape and chewing gum, I seriously doubted that the malodorous vehicle could make the 12-hour journey. The seats were appallingly uncomfortable, having no cushions, and the only air conditioning was via the open windows. It turned out that I was right about the bus as it broke down about three hours into the trip. We waited for an hour and a half for a replacement vehicle to arrive. Thankfully this was a little more comfortable, but it didn't stop this bus also breaking down for a while, delaying us for at least another hour. What should have taken 12 hours took closer to 16. On the way I did manage to while away some hours asleep. Bizarrely I dreamt that I was taking a long journey on a dilapidated bus with uncomfortable seats!

At one point we stopped for breakfast in a nice café at the side of a jungle road. As most of our trip was to take place on the grasslands, this was the only rainforest we saw in Guyana. Unlike most other South American countries, Guyana is taking steps to preserve its rainforests. At the UN Climate Conference in Bali in November 2007, the President of Guyana, Bharrat Jagdeo, offered to lease the UK all of Guyana's remaining rainforest. That is a total of 20 million hectares. The trees would be unharmed in return for aid from the UK. The government is currently considering this remarkable offer.

We drove through farmlands, deep jungle, and finally on to the grasslands. We caught a ferry across the Essequeibo River, the largest in Guyana, and Damon told us that there were islands in the wider parts that were larger than Barbados! Whilst on the ferry, we met a pleasant American girl called Rhiannon. She told us that she worked for the VSO and that she had been in the country for many months.

We eventually arrived in Letham, and took an open-backed truck out on to the savannah. On the way, we saw many birds such as caracara, egrets and jabiru. The landscape was very sparse of trees, but scattered with termite mounds that looked for all the world like Christmas trees constructed out of mud.

We reached the tiny village of Toka, where we picked up some more guides and porters, as well as several girls who were to cook and wash for the expedition. We then started out towards the village of Taushida. Unfortunately, we were hiking at noon when the sun is at its most ferocious.

The heat on the grasslands of Guyana was quite unlike anything I had ever encountered before. In comparison, the heat of West Africa seems like a chilly winter's day. In Indonesia and Thailand there was shade, but this is a commodity that is lacking in this part of the world. The relentless heat, and lack of shade, affected me badly and I suffered from sunstroke. Several times I collapsed on the way to Taushida - the six miles seemed more like sixty. I had to take many rests, but during one of these I saw a hummingbird at close range. One guide, Joseph, told me that a di-di had been seen in the area. It had resembled a huge white man covered with hair, and had been seen in the mountains, peering through some vegetation.

When we finally arrived, I was able to wash in a stream close to the little village. The cool water was a blessed relief. We relaxed as tiny cyprinid fish, probably bucktoothed tetra (*Exodon paradoxus*) nibbled on our toes. That night, as we made camp in the village, a bush fire sprang up on the far side of the creek. It resembled a serpent of fire as it grew like some medieval salamander uncoiling in the night. Thankfully the flames did not reach over the water to menace our camp.

Damon told us that, a couple of weeks before, he had seen some strange lights in the sky above the mountains. This light resembled a bright star and had appeared to seemingly break into several smaller lights. They had remained visible for some time. Could the place be a window area? As I pondered this question, a meteor flashed across the sky - which was untroubled by light pollution - and I watched as it burned up in the atmosphere.

Unfortunately, we were told that Corona Falls was a full 70 miles away as the crow flies!. It would mean walking 20 miles per day, there and back, and as six miles had almost killed us, there was no way we could walk overland in the heat. We considered renting a helicopter when we got back to Letham, but, in the meantime, there was much to see in the immediate area. We were told that there was a cave at the top of a nearby mountain that contained an urn with human remains in it. No outsiders had yet seen it.

In the morning, we arose early to climb up the nameless mountain. The range, as a whole, were known as the Taushida Mountains. Five years earlier, a hunter named Moses Issac had stumbled across an amazing discovery in a tiny cave atop the mountain. Moses had found an Amerindian urn burial undisturbed from the time it had first been laid there. In the cooler morning climate, the climb up was relatively easy, but as we neared the top disaster struck. Lisa, who had been following in the path of a much smaller, lighter Amerindian girl, trod on, what she thought, was solid ground. It wasn't. She fell badly, breaking her thumb, shoulder and the soul of her foot. It looked as if the thumb was broken, as it swelled up like a little sausage. She didn't let the fall deter her, however, and climbed gamely on to the summit. She was lucky she only fell 17ft and not 300.

The remains were located in a shallow cave, and the scattering of flat rocks suggested that the entrance might once have been covered. The remains were in a large earthenware pot, and there was a whole skull of a boy aged between 9 and 12, as well as the jaws and ribs of an adult man. The ribs and jawbones were in a smaller container within the large pot. There were also small beads and the tooth of a peccary. The tooth had a hole in it suggesting that it may have once been a necklace.

Damon did not know exactly how old the remains were. They could have been pre-Columbian, over 500 years old, or as recent as the end of the Amerindian wars around 100 years ago.

Lisa was allowed to take some pot fragments and beads for analysis at Oxford University. Damon left some new beads as a replacement, and we all took a drag on a cigarette (despite the fact that everyone except Chris was a non-smoker) as this was a custom to show respect to the dead.

The older man was obviously someone important in order for his remains to have been buried in such a prominent place. Perhaps he was a shaman or a chief? The boy may have been a sacrifice. Maybe the peccary tooth was the boys first kill as a hunter? The long and short of it is that we just don't know. Perhaps the analysis of the pot fragments and beads will shed more light on the mystery.

Moses related that about ten years ago, a di-di had been seen walking across the mountain. Damon also said that in the past what looked like huge stone clubs and shields had been found. They were far too big to be used by humans and he wondered if they were made by di-di. I thought it was more likely they were symbolic objects, never meant to be used as weapons.

Another guide, a local hunter named Kenard Davis, told us a story that his father had related to him. It happened in the 1950s, when a man had been out hunting and was coming home over the mountains. The mountain pass was quicker than walking all the way around. He was holding two wild fowl he had killed, one in each hand. As he neared the top of the mountain, he looked up and saw a huge hairy man asleep in the trees. He seemed to be using the vines like a hammock.

The man was so frightened that he ran all the way to the bottom of the mountains, still clutching the birds. When he returned to his village, he fell ill and believed that the di-di had put a spell on him. He consulted a shaman who went into a trance to contact the di-di. The creature told him that the man had frightened himself into sickness. The di-di lived on the mountain, and had a wife and daughter who lived on neighbouring mountains. They did not harm people.

This kind of intense fear has been noted in other encounters with 'ape-men' around the world. Australian cryptozoologists, Tony Healy and Paul Cropper, note this in their book *'Out of the Shadows: The Mystery Animals of Australia'*. In one case, a group of young men were reduced literally to tears by the fear they felt on encountering a yowie in the countryside outside of Canberra. Similar cases have been noted by Janet and Colin Bord in their *'Bigfoot Casebook'*.

Kenard had never seen a di-di himself, but he did tell us of one strange creature he had encountered. Up until the 1970s, a tiny, red-faced pigmy was well known in the area. He was hairless, naked, brown skinned and about 3 to 3.5 feet tall. He had a red painted face and always wore a strange grin. He would leap out of the bushes, grinning at passers-by and scaring them, though he never did anyone any harm. Kenard's uncle had a motorbike, and the little red-faced man would often hop on to the back and catch a ride. He always leapt off at the same spot, which Kenard's uncle assumed was his home. According to Damon, people left gifts of tobacco out for him.

The food on the expedition was generally good. We dined on chicken, rice, fish and cassava. The latter is a root also know as the yuca (*Manihot esculenta*) which is a major source of carbohydrates. The native peoples shred it, then soak it to remove the toxins. It is then squeezed through a wickerwork tube, before being dried and pounded into a granulated form. Cassava is remarkably filling and a small portion can keep you going for a whole day. It can be eaten in a soft form that is akin to couscous, or in a hard granular form that is not unlike granola. In both of these forms, it is quite palatable. However, when turned into cassava bread it has the taste and texture of chipboard.

Later that evening, as the sun became less fierce, we travelled in the opposite direction to visit Tebang's Rock. This is a 30 foot, tall pillar of rock that stands on the savannah. Kenard told us of Tebang. He was a little man who walked around at night touching children in order to transmit disease. Once the child had succumbed, Tebang would fashion a flute from their bones, and play it atop his rock. He was still supposed to be seen on moonlit nights, whistling and shrieking. He seemed to be totally different from the red-faced pygmy Kenard had mentioned previously. Tebang is reminiscent of the African goblin, Tokoloshe; a horrid creature with an out-sized head that wanders the night transmitting illness to children by touching them.

There is another mysterious mountain in the area. Damon was told by a man, who was climbing it, that he was almost sucked into a cave by a dragon near the summit. He did not see the dragon, but the great sucking force that came from the cave's mouth had almost pulled him in. Damon thought that the force might have been wind blowing through the mountain if it was hollow, or had a network of caves. The cave might have acted like a wind tunnel under the right conditions. Where the idea of a dragon came from is unknown; there are Chinese immigrants in Guyana, but they are few in number. Could their culture have reached these remote areas? As I have noted elsewhere, the dragon is a universal monster and cannot by written off lightly as a mere myth. As the dragon mountain was ten times higher than the mountain where we had seen the burial, we decided not to climb it in the current climate!

The next day we returned to Toka. The walk went well until the sun reached its zenith, then sunstroke began to kick in again. Once we reached Toka, we all slept in the shade for several hours. On all of our other expeditions we tended to rise at about 7 am, and then trek through the day until about 7 pm. In Guyana, we could not do this on account of the heat - we were forced to stay as still as we could for the middle portion of the day, and this, of course, meant that much valuable time was lost.

Damon pointed out some cashew fruit. They grow directly below the nut. The nut is poisonous until roasted, and in the west we see only the nut and never the fruit. The bright yellow fruits looked a lot like pepper, but tasted like cranberry - juicy but leaving a paradoxically dry aftertaste not unlike a more intense version of cranberries.

We caught a small open-backed truck, and travelled a few miles down a dusty road, before walking to our next destination - Crane Pond. Here there were anacondas. Not monsters like the one at Corona Falls, but average-sized specimens. Still, it would be interesting to see one. It was cooler now, and as we walked across the savannah we came across a female giant anteater (*Myrmecophaga tridactyla)* with a baby upon her back. The spectacular beast began to lollop away from us, until Damon chased and drove it back towards us, allowing us to film it, and we got some wonderful film of her before she galloped over the horizon. It was a dangerous endeavour as the anteater is armed with formidable claws that it uses, not only to rip open termite mounds and anthills, but in defence against predators. Anteaters have also been known to disembowel humans who attack them. In fact, in **April 2007, an anteater at the Florencio Varela Zoo in Argentina attacked Melisa Casco, a 19-year-old keeper, fatally clawing her.**

As the sun began to fall, we set up our tents. Kenard told me that Crane Pond was once thought of as the lair of a dragon. There was an old saying that ran: 'don't sleep too deeply at Crane Pond or the dragon will take you'. He said that, back in the 1950s, when cattle ranching was still big business in Guyana, a group of cowboys had camped at Crane Pond for the night. During the night they had heard a huge animal rising from the water and could hear its breathing. The cowboys had panicked, some gathering up their horses, whilst others fruitlessly fired their rifles at the noise. They beat a hasty retreat.

Could the dragon have been a giant anaconda? Anacondas do make a strange sound when breathing, which has been likened to snoring.

That night, I was badly bitten by insects. In the cool morning before sunrise we awoke, and at first light set out in search of anaconda. We found large furrows filled with water going to and from the swamp. They were the trails of anaconda, and by the look of them they would have been 15-17 feet long, which is the average size. Kenard saw a baby anaconda of around 4 feet slip into the water, but the larger ones eluded us.

Looking back now, I wished that I had waded deep into the middle of the pond barefoot feeling for anaconda with my toes. This is the way ordinary sized ones are captured by biologists in places like Venezuela. But the savage, unyielding, merciless heat once more began to breath down like a fist of lava.

We followed a line of trees along a partially dried-up creek, but found no anaconda. As the sun was getting higher and hotter all the time, we decided to return to camp. Kenard travelled further down the creek, hunting for game to supplement our rations. He returned, dragging something through the long grass. I thought he may have shot a bird with his bow and arrow, or even a young capybara. He had, in fact, killed a small spectacled caiman (*Caiman crocodilus*). I have always said that I would never eat crocodilian meat, as I am so fond of them, but it would have been boorish to refuse something Kenard had killed specifically for us to eat.

The spectacled caiman is in no danger, in fact it is one of the commonest and most widespread crocodilians, and is sometimes hunted for food. Kenard had made a clean kill - shooting it through the skull with an iron tipped arrow. It made you appreciate just how tough crocodilians are, as the two-inch iron arrowhead had been bent right round by the caiman's hard hide and bone.

As the sun got hotter, we returned to camp. Jon and Chris decided to walk with Kenard to a ranch a few miles away to get some pop and water. We had all been drinking out of flasks that purified water from streams and ponds. The thought of pop was appealing, but I also thought they were both insane for wanting to walk in the blazing noon sun.

Back in 2004, on our second expedition in search of orang-pendek in Sumatra, Jon had gone off on an equally odd quest. Whilst on route to an unexplored valley in Gunung Tuju, we had stopped overnight at a local farmhouse. At around 9 am Jon announced he had a craving for pop. Our guide, Sahar, offered to walk back with Jon to a remote trading post we had passed hours before. The two walked off into the dark and were gone for over four hours. We were getting quite worried by the time they finally returned. The pop that they had brought back was a delightful local brew called 'Frambosen'. It was put into old 'Fanta' bottles and it was obvious that the manufacturers had tried to make the 'Frambosen' fizzy. But they had got the mix wrong, and it was just frothy like soap suds. It tasted like old, flat, warm 'Vimto'. It was not, in my opinion worth the four-hour trek to get it.

After they left, the heat began to rise. Even the guides said it was remarkably hot. The sun was truly unbearable. There was no respite in the tents, as they merely magnified the already savage temperature. We poured water over our heads, but that only elevated the heat for a short while. Finally, in desperation, I waded waist deep into a swamp and stood under a tree for several hours. The mosquitoes were nothing, a minor irritation in comparison with the solar torture I was enduring. I have always preferred the warmth to the cold. I have always favoured summer over winter, and sun over rain, but on that day I prayed for rain, or even the slightest cooling breeze. Now I have felt the wrath of Guyana's dry season, I will never feel quite the same about cold weather.

Paul also began to suffer badly from the heat at this point. He later almost blacked out whilst packing up his tent.

Jon, Chris and Kenard finally returned with pop and water. It was a nice change to be able to gulp down liquid, rather than having to suck it up from a water purifying flask.

That evening we moved on to Cashew Pond in an attempt to avoid the insects. The walk was supposed to have been 2 km, but it was probably closer to 4 miles (6.4km). The terrain was very bad. The uneven ground was dotted with concrete-hard lumps that were hidden by the dry grass. Lisa was suffering with her injured foot, and she now had blisters upon blisters, and it was beginning to look as if she might lose a toenail. Her injured shoulder also meant that carrying her heavy pack was very painful.

That night we roasted the caiman on an open fire. The best meat was to be found in the legs and tail - the legs tasted like chicken, whilst the meat of the tail tasted like very flavorsome, chewy cod. Once again, the insects made a meal out of me, and, in fact, we were all bitten worse here than at Crane Pond.

In the morning, we packed up to move out to Point Ranch, where we were to be picked up by mini bus and taken back to Letham. The walk was long, hot and uncomfortable, especially for Lisa whose injuries were really hurting her.

Whilst we waited at Point Ranch, I asked about the water tiger. An old man called Elmo, who came from the ranch, had seen them. He was adamant that they were not the giant otter *(Pteronura brasiliensis)* with which he was familiar. Elmo said that the water tigers he had seen were spotted like a jaguar *(Panthera onca)* but hunted in a pack. He said that there was a 'master', possibly a parent, that sent out the cubs ahead of it in order to flush out prey. He had seen a whole group of them several years ago. Elmo pointed out a local mountain where he said that a pack of water tigers lived. The mountain had no name, but it was said that a dragon guarded a spring there, and Elmo added that no-one who had ever climbed it had returned.

Kenard confirmed that water tiger was supposed to come in different colours; spotted like a jaguar, brown, or white with dark spots.

Another of our guides, Joseph, stated that he had seen the hide of a water tiger killed by a hunter in the 1970s. It was 10 feet long, including the long tail. It was white (he compared the shade with some cows on the ranch) and had black spots. The head was still attached, and he said that it was striped like a tiger.

These descriptions, both physical and of their behaviour, match no known cat species. Lisa thought that the water tiger might actually be a form of giant mustelid, as certain species such as stoats can change the colour of their coats.

Joseph had another, even stranger, story to tell. In 1975, a plane had crashed into a mountain in the same range as the one supposedly inhabited by a dragon and a family of water tigers. He had been paid to climb up and retrieve the body of the pilot from the crash site. Joseph found the wreckage and the corpse, but it was missing its head and had been badly burnt. He retrieved the body, and put it into a sling fashioned from a blanket and began to descend, but on the way down he became hopelessly lost. He wandered for three days on the mountain before finding his way down, and during this time he was forced to consume the flesh of the crash victim, in order to survive, and he had eaten one of the arms.

Joseph had been reticent to tell us of his homophagy, worried that we would think he was a cannibal. But we all agreed that we would have done the same if it were us.

We took a truck back to Letham. We had decided that none of us could stand another 18 hour plus journey in a rickety bus, so we opted to fly back to Georgetown when we left. We booked our tickets in advance at the little airstrip in Letham, and whilst there, we enquired about chartering a helicopter. Unfortunately, there was only one in the whole of Guyana, and it was not available. We considered chartering one from Brazil, but this would have meant days of red tape. We also considered a boat, but Kenard said that the river was too low. All of this was amazingly frustrating, as the main thrust of the expedition was to search for a giant anaconda at Corona Falls, and we had not even seen one, solitary, ordinary-sized anaconda so far. The fact that the pool was unreachable was like a dangling the proverbial carrot before An equally proverbial donkey. It looked like we would not be reaching our target on this trip, and we decided to try and return next year, during the rainy season, and charter a boat or plane to get to Corona Falls. There was an airstrip only half an hour's walk from the pool, but no aircraft available presently.

Both Kenard and Damon mentioned that, during the airstrip's construction a number of years ago, eleven skeletons had been found inside termite mounds. The skeletons were in crouched positions, suggesting that people had broken open the termite mounds, placed the bodies inside them, leaving the termites to rebuild their mounds around the cadavers.

Neither Damon nor Kenard knew of any tradition in any Amerindian tribes that had funerary rights like this, and Damon postulated that it could have been pre-historic. The bones had, unfortunately, been thrown away, and no further research had been carried out.

We checked into a guest house again, and that night a herd of horses stampeded through the hotel grounds. None of the locals batted an eyelid! The next day Damon had arranged for us to meet with a former tribal chief who knew a lot about the strange creatures of Guyana.

We drove until the savannah changed to jungle, and then we drove up a twisting jungle path to a clearing near a stream. Waiting for us there, was a middle-aged man in a 'Sideshow Bob' t-shirt, and holding a parang. He introduced himself as Ernest, and told us that he had been a tribal chief until about eight years ago, when he retired to concentrate on running a little fish farm at the base of the Kanaku Mountains.

Ernest was a wealth of information on all of Guyana's monsters, and then some. About ten years ago, and around thirty miles away he had seen a 30 foot anaconda in a pool. He said that it had been shot by an Englishman, and that the skin had been transported to England. This, if it was imported, would have been done illegally. He knew of the di-di, but had not seen one himself. However, a friend of his, who had died two years previously, had seen a di-di. His friend had seen a female suckling an infant in a tree, and had watched them for a while before blacking out. Afterwards, he fell ill and as his illness became worse and worse, he blamed it on seeing the di-di. He only admitted to the sighting on his deathbed. This reminded me of our 2006 expedition to The Gambia in search of the dragon-like Ninki-Nanka. Local people there believed that to see the Ninki-Nanka was death - you would die within five years of seeing it. Ernest said that the voice of the di-di, like a very loud human shout, was still heard from time to time in the Kanaku Mountains. He said the di-di live high up in the mountains, and only come down to lower levels to hunt.

Ernest knew of the little red-faced pygmies. When he was 19 (he is now 59) he had seen one. It was naked, brown-skinned and had a red face. Unlike Kenard, he felt that the red face was natural pigment, and not painted on. The little man had taken tobacco from Ernest before vanishing back into the forest. He told us that the pygmies are more often seen than heard, that they liked tobacco, and were not dangerous unless angered. They sometimes made homes under large trees, and if one of these were cut down, the pygmies, quite naturally, would get angry. They also made little pots that humans sometimes came across in the forest, and these, too, should be left alone. The pygmies did not speak to humans, even when spoken to - they seemed just to take tobacco and leave. He called the little men 'bush people'. He, unlike most others, thought that the red face was natural and not painted on. He said that they were chubby and wore no clothes. They had hairless brown skin.

Damon confessed that he had seen a pygmy as well. About ten years ago he had been camping with his sister-in-law, and another girl. He awoke in his tent to see a tiny red-faced man grinning down at him. He was frozen with fear, until finally he found he could move enough to try to nudge the girls awake. When he looked again the pygmy had gone. He did not recall hearing the zip on his tent being pulled open.

20 years ago, Ernest had a run-in with the water tiger. He and his grandfather were on a small boat on the river, when something seized the vessel from beneath, and started to shake it. Ernest and his uncle had to hold on to some branches overhanging the river in order to stop the boat overturning. Ernest's grandfather said it was a water tiger, though neither man saw the attacker. It could just as easily have been a big caiman. He, too, said that the water tiger lived in rivers, and ran in packs. He said the water tiger could be brown, spotted like a jaguar, or white with black spots. The tail was flattened like an otter's tail and the head was like a jaguar. The water tiger's paws were thicker than a jaguar's paws.

Ernest's final story was of something of which none of us had previously heard. A couple of years ago, in a little cave at a place called Wa-sa-roo, he had seen a tiny caiman. It was smaller even than the smallest known species, Cuvier's dwarf caiman (*Paleosuchus palpebrosus*). It was brown in colour, and had a red strip running down its back. The description matches no known caiman species, but stranger than this, he said that the tiny caiman had two tails!

Lizards and snakes have sometimes thrown up freak specimens with two tails due to genetic deformity. However, as far as I am aware, this has never been recorded in crocodillians. Could Ernest have seen a pair of caiman mating, one on top of the other? He did say that the tiny animal was making a very loud bellowing noise, out of proportion to its modest size. Male alligators are known to bellow loudly during the mating season, so this seems like a reasonable explanation. Ernest had also seen the little caiman in a creek near the cave.

We thanked Ernest, and set out to explore Wa-sa-roo. It was a collection of boulders, some of which were house-size, through which a stream ran. I took off my shoes and scrambled into the small cave. It was cool, and had water and ledges. Though there was no evidence in the form of tracks etc., the cave was the perfect place for a small caiman to make its lair. Dwarf caimans like fairly fast flowing, rocky streams, and Guyana could be playing host to a new, unrecorded species.

We then climbed up on to the top of the boulders to look down into the caves. Lisa used the night vision setting on her camera to take film of the inside of the caves.

Later, back at the guest house, Kenard told us some more stories that he had heard of the di-di. Once, many years ago, a hunter found a huge, human-like footprint miles from any habitation. He followed the tracks till they came to a tree, and looking up he saw a huge hairy man sleeping in the vines. He ran away in fear.

In the 1940s, a girl was kidnapped by a di-di and taken deep into the jungle. It took her as its mate, and together they had a hybrid child - half-human, half di-di. The girl stayed with the di-di against her will, until one day she saw a hunter in a canoe. She shouted him over to the bank and leapt aboard the canoe. As the hunter paddled off, the di-di emerged from the jungle and stood on the bank gesticulating for the girl to return. When she did not, the monster picked up their half-breed offspring and tore it to shreds like a doll.

Quite where this story had its genesis is an interesting question. I have heard exactly the same story told about both the yeti and the sasquatch.

Another story told of a group of men boating down the Essequeibo River. At one point, they had stopped and disembarked. The men saw one of their colleagues grabbed from behind by a massive hairy arm, and a huge ape-like figure carried him into the jungle. The other men pursued, shooting at the shaggy giant till it let the man drop.

That night, we were invited to dine with Ernest's family at their home just outside of Letham. During the meal, Damon mentioned that several members of his family had once worked in the largest open cast gold mine in Guyana. It was once owned by a Canadian company, but has since been sold. However, they had employed many Amerindians. One time a whole village full of people witnessed the uncovering of a huge, human-like skull. It was far larger than a man's skull, and Damon wondered if it could belong to *Gigantopithecus blacki*, a giant Asian ape believed extinct for 50,000 years. Many think that the yeti and sasquatch may be a surviving form of this ape, or something related to it. So far *Gigantopithecus* remains have only been found in China and India. Officials from the company owning the mine came and took the skull, and it was never heard of again. Perhaps the company was scared of having the mine closed down if a major paleontological discovery was made. Maybe this skull is locked in some bureaucrat's basement to this day.

Foster recalled an even odder story of some man-like creature with webbed digits being swept into a village during a flood. It was described as resembling 'the Creature from the Black Lagoon'. I too have a vague memory of hearing something like this many years ago. Just what the creature was, if it ever existed, and what had become of it, was unknown. Details were lacking, but Damon was sure it had appeared in the local paper many years ago. There was an on-line archive of the paper, and Paul said that he would check it out when we got back to England. I looked at the online archives for both of the country's papers when I got home. Neither the *Guyana Chronicle* or *Stabroek News* had online archives older than a few years. Nothing in these referred to the story of the strange creature. I have e-mailed the editors of both newspapers asking if their archives go back any further, and if they can recall the story. Rest assured that if I find anything, it will be appearing in the pages of *Animals & Men*. There is a tradition that is widespread in South America of small aquatic beings known as 'Negroes of the Water'. If such aquatic goblins, about whom there is little information, are based on some real creature remains to be seen.

The following day, we caught the tiny plane back to Georgetown. Damon and Kenard had to stay in Letham as they had to pick up some snakes and invertebrates from remote villages.

Whilst waiting for the plane, we again met Rhiannon, the American girl we had seen on the ferry. She told us that she had been researching a kind of spirit /creature called

a Kanima. It seemed to be a different thing to each tribe. One said it took the form of a short fat man.

Foster said he had spotted a man who knew all about the Kanima and went over to fetch him. He was a rubber tapper by trade, and fashioned his rubber into nice model birds that he sold to shops. He told us that the Kanima was a sort of solitary witch doctor, a human with magickal powers. Kanima lived alone and would pass on their knowledge to suitable students who sort them out. Only men were Kanima. Kanima could lay curses or cast spells but they usually left you alone unless you provoked them. One power he mentioned was to use certain leaves to become invisible in the forest. This may sound unbelievable, but it might be a matter of what is meant by 'invisible'. Tribesmen in Peru have long said that the secretions of the giant monkey tree frog (*Phyllomedusa bicolor*) made men invisible, gave them great stamina, and allowed them to go without food. Recently biochemists analyzing the frogs' secretions have found powerful laxatives, diuretics and emetics that may well flush out smelly compounds from human skin, making the hunter invisible to forest animals that mainly rely on their sense of smell. The secretions seem to have painkillers and hunger suppressants in them as well. Could the plants used by the Kanimas have similar properties?

The rubber tapper also said that, as a boy, he had seen a red-faced pygmy. He had been hunting in the general area, and had seen a tiny man with a red face peering at him through the undergrowth. Unlike other witnesses, he thought the man was hairy (though this may have been an animal hide he was wearing). When he realised he had been seen, the figure fell to all fours and ran off. This is the only mention of hair and moving on all fours. Could this witness have mistaken a monkey? The red faced black spider monkey *(Ateles paniscus)* fits the description. It is hairy, has a red face, and moves on all fours, but can also stand erect. However, it has a long tail and inhabits rainforests, not the dry savannah around Letham.

The rubber tapper also confirmed the story of the burials in the termite mounds.

When we caught the plane, Paul was nervous. He had never been in such a small aircraft and described it as a tin shed on wings. During take off, the back of the pilot's chair sheered off and fell into Paul's lap. This did nothing to allay his fears.

On reaching Georgetown, Damon's nephew Marradonna was waiting for us. He was to be our guide for the last few days. Paul wanted to see if he could get his flight brought forward. He was missing his children and had had enough of camping, heatstroke and insect bites. We took him to the main airport, and he found out that he could get to Barbados, where we were due to transfer to Gatwick, but they did not know if he could then get his Gatwick flight brought forward. He decided to risk it. The flight was not until 3am, so he arranged to stay with our taxi driver, a chap called Marvin, until then.

The rest of us headed back to Pakuri. As it transpired, Marvin's house was a malodorous shanty with no glass in the windows and delightfully situated next to an open sewer. The inside was crawling with vermin, so Paul decided to accompany Marvin on his rounds. He later told us that the whole city seemed tense with the air of violence about to explode at any moment. At one point the van was surrounded by screaming men, who were all banging on the windows. Marvin was an Amerindian, but had turned his back on the culture. For some insane reason he preferred to live in the horror of Georgetown, than in the peace of a native village.

Back in Pakuri, I spoke to Foster's father Joseph. He had heard of the di-di, but had never seen one. He told me that before Pakuri was built, the area was well known for the little red-faced men. They would want to wrestle with humans and were very strong. The way to supposedly defeat them was to knock them over, as their legs did not bend and they could not get back up again. This was a story we had heard told about the di-di, that their legs cannot bend. It has also been told about the Asian yeti. Quite where such an odd notion comes from is a mystery. No mountain-dwelling animal would get very far on legs that didn't bend. In fact, immobile legs would be a non-starter for almost every animal.

Joseph had seen a large anaconda once. It had been flushed into the village by floods about five years ago, and was 23 feet long. His father had told him that a water tiger once lived in the area, where it had a lair in a cave by the riverbank. It had brown fur, and was supposed to be dangerous.

We bathed in the red waters of the creek once more, and later we were taken on a canoe ride for several miles. We saw little wildlife however.

We slept in Joseph's house. That night we watched Guyanese television (the village had recently got a generator). There was a programme about racial unity in Guyana. There is a lot of tension between the blacks and the Asians, but no mention whatsoever was made of the Amerindians. Amerindians have almost no rights in Guyana, and, despite being the native people, are looked down on by blacks and Asians. These two groups have all the power in the country, and the native peoples have none. They mainly live inland and have made far better communities on their own. Georgetown is rife with crime and violence, but Pakuri is a wonderful community, free of theft and violence.

Marradonna told us of how Damon was breeding the endangered red footed tortoise *(Geochelone carbonaria)* on Barbados. The species is swiftly dying out in Guyana, and he wants to import his captive bred tortoises to boost the wild population, but the Guyanese government will not let him. They have offered no explanation, and probably merely hold him in contempt because he is a native.

We travelled back to Georgetown, and checked into a surprisingly pleasant guesthouse. It had a colonial feel, and was situated in the only nice area of Georgetown, not far from the Prime Minister's house.

We decided to visit Georgetown Zoo, but this turned out to be the low point of the whole trip. I'm a former zookeeper and a supporter of good, responsible zoos. Georgetown Zoo was one of the most appalling collection I have ever witnessed. Green anaconda *(Eunectes murinus)* and common boa *(Boa constrictor)* were in enclosures devoid of water. The anaconda is a semi aquatic species and requires access to water at all times.

A pool full of Schneider's dwarf caiman *(Paleosuchus trigonatus)* were being kept in filthy, over-crowded water with little basking room. One specimen had had its top jaw bitten clean off. A jaguar was being kept in a tiny cage, and was only released out of its sleeping quarters until about 1.30 am. A brown capuchin monkey *(Cebus apella)* was stuffed into a tiny bird cage with no food or water. The water in the lake that supposedly housed manatee *(Trichechus inunguis)* was filthy, and despite the 'no-fishing' signs, about twenty men were happily casting away in the lake. Worst of all, was a full grown African lioness *(Panthera leo)* that was crammed into a 12 foot by 7 foot bare concrete enclosure. The poor animal had sores on its hind pads, was exhibiting stereotype, repetitive behaviour, and was being driven mad by its sterile, tiny enclosure. The keepers seemed to do nothing, but lie around in the sun.

I am currently tryng to contact various animal rescue groups, including the Born Free Foundation, to try to get these animals moved to better accommodation. I would strongly urge all readers to write to the Government of Guyana, and complain about this sickening excuse for a zoo.

The next day we visited two museums. The Walter Roth Museum of Anthropolgy[**]

Walter Edmund Roth (2 April 1861 – 5 April 1933) was an English anthropologist and physician, active in Australia. In 1906 Roth was made protector of Indians in the Pomeroon district of British Guiana. In 1924 his valuable *An Introductory Study of the Arts, Crafts, and Customs of the Guiana Indians* was published. Though called an introductory study this is an elaborate work of well over 300,000 words with hundreds of illustrations. A volume of Additional Studies of the Arts, Crafts, and Customs of the Guiana Indians was published in 1929 as *Bulletin No. 91* of the Bureau of American Ethnology. Roth retired from the government service in 1928, and became curator of the Georgetown museum of the Royal Agricultural and Commercial Society, and government archivist.

was poorly laid out and no dates were given for any of the artefacts. Lisa, an archeologist herself, asked the resident archeologist if he had heard of the crouch burials in the termite mounds. He dismissively snorted that 'he had never heard of them'. We told him that a number of locals had told us the same story. He asked who our guide was, and when we told him he laughed derisively and said *"Pha, Damon Corrie, you don't want to listen to what he tells you. His information is wrong. He is from Barbados, he is not even from Guyana."*

In fact, Damon is a native, a Chief of the Eagle Clan Arawaks. We had been to his village and had met his relatives. Lisa became angry and walked out. I told the man that Damon was an Amerindian and had shown us things that that the pompous curator had never seen, or heard of. I also told him what I thought of him, and his pathetic excuse for a museum. If you are ever in Georgetown, don't bother to go to the Walter Roth Museum, it's a waste of time.

The National Museum was a queer beast! Downstairs was an odd collection of tacky souvenir-type dolls, such as you might find in a tat shop in a touristy area of London, including a Scottish piper. There was also a collection of toy cars. Next to these was a case full of totally unconnected objects including a box for holding string, and a knuckle duster! The next display was a collection of very poor paintings and sculptures. These included the bust of a Health Minister for tuberculosis, who died in Nottingham in 1952. It looked as if it had been made by a retarded school boy.

Upstairs was a collection of badly stuffed animals. Most were in a very poor state of repair, and had been bleached due to being left in direct sunlight. A paca (*Agouti paca*) had been so badly stuffed that it looked like Tove Jansson's Moomintroll! They had a moderately-large stuffed anaconda that I managed to film despite the museum attendant's bleating protestations. All in all, the zoo, and both the museums, were run in true 'Georgetown style' (that is appallingly badly).

It was a relief to get out of Georgetown. We had ten hours to kill in Barbados, and I could think of far worse places to be than there. We were met by Damon Corrie's charming father who looked just like him, and - to our surprise - Paul Rose! Paul had not been able to get a linking flight, and *ergo* had to suffer three days in the blue seas and white sands of Barbados! We all felt sorry for him after this undoubted ordeal.

We were taken back to the Corrie residence to meet Damon's mother, brother, wife, and children. A strong resemblence ran through all male members of the Corrie family. On his mother's side Damon was descended from an Arawak princess. Her son had handed over the cheifdom to Damon upon his death in 1998. Damon could also trace his relatives back to Scotland in the 10th Century!

After lunch, Mr Corrie had arranged a tour of the islands for us by mini-bus. Barbados really does live up to its reputation as a tropical paradise. After the depressing filth of Georgetown, it was a breath of fresh air. Barbados has the bluest and most inviting seas I have ever seen, but sadly we did not have time to swim in them.

The first stop on our tour was a fortean surprise - the Chase Tomb of the 'creeping coffins' fame. This was a place I had always wated to visit. The tomb stands at the entrance to the Christ Church Graveyard and is built of large cemented blocks of coral. It measures 12 feet by 6 feet and is sunk halfway into the ground.

Nothing happened for the first two burials on 31st July 1807 (Mrs Thomasina Goddard) and 22 February 1808 (an infant, Mary Anna Maria Chase). However, on 6 July 1812, the tomb was opened to bury Dorcas Chase. It took several men to open the heavy door and they found that the two coffins already there had been flung against the wall. As both coffins were encased in lead, a great force was needed to do this.

There was no disturbance in the dust on the tomb's floor. They buried Dorcas Chase, and returned the other two coffins to their original positions.

On subsequent burials, the same thing occurred. Each time the coffins were found flung against the walls. At the burial of Thomasina Clarke, which took place on 17 July 1819, the Governor of Barbados, Viscount Combermere, supervised the sealing of the vault. Nine months later he returned to check the state of the tomb, and again found it in disarray, yet the seals on the door that he had personally put there remained intact.

In 1820, the vault was emptied without the mystery of the "creeping coffins" being solved. The coffins were all reburied in another place, and there has never been a satisfactory explanation for the creeping coffins. If they were moved by floodwater, then why were neighbouring tombs not affected? If it was animal, why were the remains not strewn around the tomb? If it was a human agency, why was there no sign of the coffins being dragged or lifted?

We were also shown historic sugar cane plantations, the remnants of a railway, and the ruins of a once, grand mansion that burned down during the shooting of the film *Island in The Sun*.

We returned to the Corrie house for an excellent meal of flying fish, roast cassava (not unlike potato in this form) and salad. The Corries were truly hospitable. Mr Corrie would not even let us pay for the tour.

Finally, we had to leave to catch the plane back to cold, dark, wet England. We all wished we could spend more time on Barbados.

So what are my final thoughts? Not getting to Corona Falls, and the giant anaconda lair was a blow. We intend to rectify this in the near future. With a small target area, a return expedition to Corona Falls would be quite easy to do. On the upside, we did turn up fascinating information on other cryptids.

The little red-faced men have never been recorded, or written about, anywhere else to my knowledge. I feel that they could be a type of tiny hominid, related perhaps to *Homo florisensis* , the tiny species of man known from 12,000 year old sub-fossil remains discovered on the Indonesian island of Flores. They still seem to be widespread, but no-one has ever studied them.

The di-di may be a bigger hominid, something related to the sasquatch perhaps. If a littlefoot can exist in Guyana, why not a bigfoot? Much of what is attached to them seems like universal folklore. They seem less common now than the pygmies, and it seems that whenever human habitation springs up, they retreat further into the wilderness.

The water tiger, despite my initial suspicions, seems to be very different from the giant otter. It is social, aggressive and comes in several colour variations. It is a flesh-eating mammal of some kind, possibly a felid or mustelid.

Lastly, the tiny caiman. This is intriguing, and could constitute a whole new species. We need to gather more information and eyewitness accounts.

Guyana is a veritable menagerie of cryptids. Few expeditions have looked for these creatures before, so the country promises to be a fertile ground for research for many years to come.

THE CENTRE FOR FORTEAN ZOOLOGY

GUYANA 2007

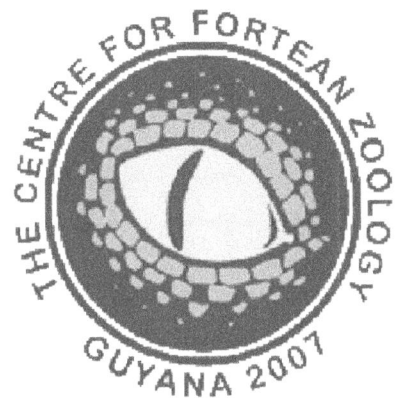

PAUL ROSE

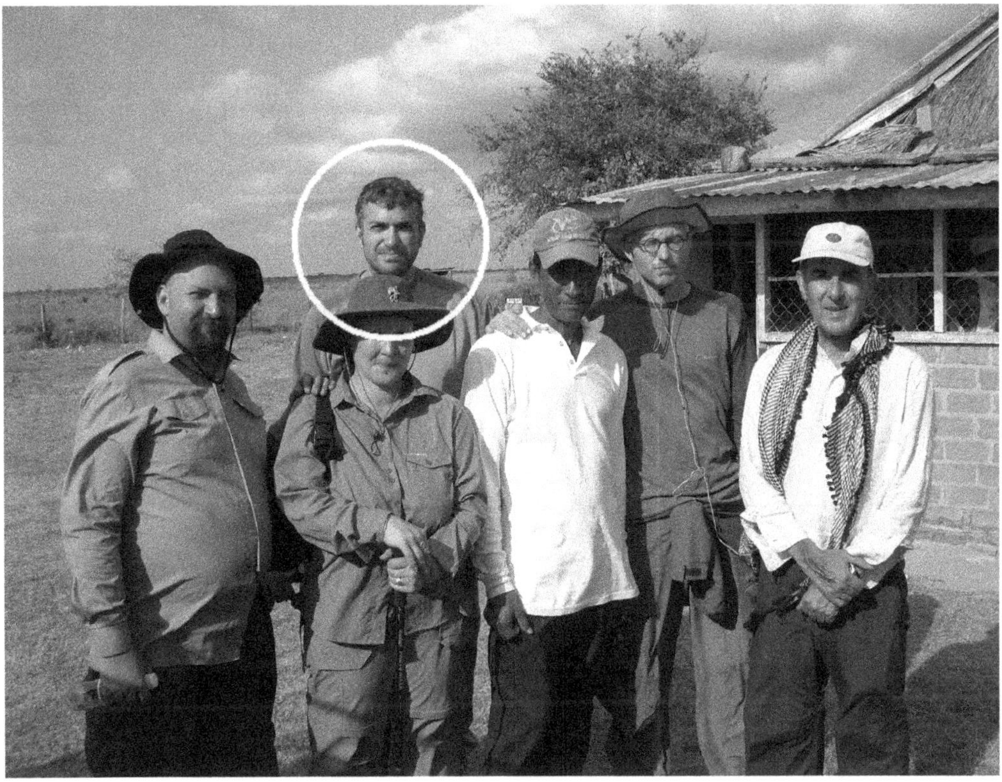

It was thirty years ago, on a particularly brittle autumn afternoon, that I had my first encounter with a monster. I was six or seven, or thereabouts, and my mother had deposited me at the home of a family friend. A woman who couldn't have been any brassier had she lived inside a giant trombone, Sheila Wignall was a peroxide chimney with a throat full of gravel and tar. She didn't so much speak as wheeze out a wall of white noise through which you could just discern a volley of dropped aitches, and a sort of rattling rasp that I now presume was her voice. The formica-and-leatherette booze bar in the corner of her living room pretty much tells you everything else you need to know. Suffice to say, I was always somewhat intimidated by this fog-enshrouded harridan, but I enjoyed my visits due to the presence of her son Paul.

Paul was a good few years older than me, and the sort of boy who stamped on frogs. He knew how to swear properly, proudly sported several ripe crops of acne, and I considered him pretty much the coolest person on the planet. Paul could always be relied upon to come up with something imaginative to play. His games were always called things like Coppers and Hooligans, Nazis and Prisoners, and Escape From Colditz (which, to be honest, wasn't that far removed from Nazis and Prisoners). Many years later I learned that Paul had developed a fondness for another game that he liked to call 'Vodka and Tonic'. Regrettably, this game ended badly for Paul when he died of liver failure. I digress.

On this particularly perfect autumn Saturday – Paul's first treble gin still just a glint in the corner of his pancreas – we spent some time out the front of the Wignalls' decaying, pre-fab home, playing with Paul's skateboard. As was usual with Paul's more exciting toys, playing with his skateboard meant I'd sit by while he entertained himself, and expected me to watch ("Look at me! Look at what I'm doing now! What's that bloody face for? I let you prod the frog, didn't I?").

On this occasion my ceaseless whining caused Paul to tell me that if I didn't shut up he would release the monster. In countless visits, this was the first time I had ever heard Paul mention the monster. Indeed, it was the first time I had ever heard anyone mention a monster as a matter-of-fact actuality, rather than as a thing in a film or story. For a seven-year-old, finding out that you're sitting just inches away from a real monster is probably the last thing you need to be told, short of hearing that Santa Claus has gone a bit funny, and decided to give everyone pleurisy for Christmas. Paul explained that all it would take to release the monster would be for him to lift up the small rock that it lived beneath. Apparently, the monster would "have a crack" at me, because I had probably already woken it up by continuing to ask for a go on the bloody skateboard, and whingeing about whether I could have my bloody handcuffs loosened a bloody bit, and anyway, it'd probably bloody just bloody have a bloody crack at me any bloody way, because it didn't like the look of my bloody stupid bloody face and bloody that. I angled myself towards the far side of the step, checked for possible escape routes, kept one eye on the monster's rock, and remained that way, more or less in silence, save for the occasional quavering intake of breath, while Paul continued to try and stay upright on his bloody skateboard.

With hindsight, I think it's highly unlikely there had been a monster living under the rock. For a start, it wouldn't have been a particularly big monster, because it wasn't a particularly big rock. I strongly suspect that the only thing under there was ordinary Yale front door key, which couldn't have done a great deal of damage even if Paul had used it to stab me repeatedly in the knee. Regardless, the memory of that day was still with me some twenty years later when I finally saw a monster for real. Except this time I was significantly less prepared to believe.

Along with other assorted members of my ludicrously extended family, my wife, kids and I had been staying on a farm on the Isle of Wight, in a mouldering caravan that had no electricity or running water (merely an abundance of running insects). One evening, we'd driven off in search of whatever passes for civilisation on the Isle of Wight, a hot meal, and respite from sharing our living space with the local, six-legged fauna. We had pulled into a lay-by on the crest of some remote hill, and were waiting for the rest of my ludicrous extended family to catch us up. As we sat, the engine tick-tick-ticking over, something stepped out from the bushes. Actually, even now I hesitate to describe it as a 'something'. I fear that word immediately implies 'something weird', in the way that saying "Oh, you're one of *those* people" is infinitely loaded. Yet it undeniably was something weird, and had I been the only one who saw it I probably wouldn't be typing this now. For over a decade, I kept this encounter more or less to myself for fear of being labelled a crackpot, or written off as the sort of person who expands perfectly ordinary incidents into the most perverse, exaggerated fantasies. Even so, my wife saw it too, and I know from bitter experience that she has no more truck with perverse fantasies than I do. Plus, I suspect I'm in good company here.

Anyhow, this thing we saw was about the size of a greyhound, yet it looked more like a cat. It had matted, reddish-brown fur, with a slender neck ending in a feline head, and a long, thin tail that stood bolt upright. Its back legs were muscular, but it was sinewy and scrawny, and we were both pretty certain that it wasn't any type of cat or dog we were familiar with. It didn't look like a domestic animal, but really – who knows? For reasons I don't really understand, I desperately wanted to believe it was a fox, yet it looked like no fox I'd ever seen (and living in North London you can barely walk out of your front door without treading on one). Admittedly, as far as most monsters go, the dog-cat-scrawny-thing didn't do anything particularly impressive. It didn't leap on the bonnet, draw back its lips to reveal four sets of needle-like teeth, and then violently smash its face through the windscreen in order to suck out our brains. It simply crossed the road not more than six feet in front of us, and – in true monster-sighting tradition – disappeared into the thicket opposite before we could snap so much as a blurry, indistinct photograph that could've been anything from a panther, to a field mouse, to Alan Titchmarsh, nude and browned-up, scurrying across the road on all fours.

The thing is, whereas 20 years previously I'd been prepared to accept a budding alcoholic's word that there had been a monster living under a pebble in suburban front garden, I no longer wanted to entertain the possibility that I'd witnessed anything out of the ordinary on that wind-scarred Isle of Wight hillside. And that was in spite of the fact I'd clearly seen something that was sufficiently freakish, that my wife and I swore aloud in front of our three young children. In the months following, I tried to hard dismiss the sighting. It was a definitely a fox, probably. Or a steroid-enhanced, starving dog. Or something else with a perfectly rational explanation. Whatever it was, I decided, it wasn't anything anywhere near as strange as it had actually appeared to be, and over time I successfully suppressed any notions to the contrary. Somehow, in the two decades between monster encounters I'd become conditioned into automatically disbelieving the unbelievable, and was no longer prepared to entertain the possibility of the extraordinary. As those months evolved into years, I began to feel conflicted. The more I thought about what I saw, and – more pertinently – the fact I'd repressed it, the more it began to trouble me. I mean, had I really become that boring? It would be a whole decade before I met someone who I felt would listen to my story, and not cough in my face, or patronise me, or insist that it was indeed just some big, eff-off fox. Enter, stage left, Mr Jonathan Downes, and the CFZ.

Back in May 2007 I had a book published under my occasional pen-name 'Mr Biffo'. Off the back of that tawdry pamphlet's release – a series of transcripts of my questionable adventures online while pretending to be a woman – I'd received an eMail from Jon, in which he was generally rather nice and gushing. While I've always wrestled with the bizarre notion that I should have what could only be classed as fans,

there was no question that this was a bit of bona-fide fan mail (in, fact, Jon says it's only the second piece of fan mail he's ever written – the other being addressed to a little known avant-garde artist called Yoko Ono). I'm always flattered when anyone takes the time to say nice things to me – I'd have to be a peculiarly hard-hearted git not to be – and I usually take time to respond with a brief email of thanks. With Jon it was different. At the bottom of his email was a link to both his blog, and the home page of something called the 'Centre for Fortean Zoology'. A few clicks later, and to say my curiosity was piqued is an understatement akin to describing the Great War as "A game of British Bulldog that got a bit out of hand".

"At the beginning of the 21st Century monsters still roam the remote, and sometimes not so remote, corners of our planet," proclaimed that alarming mission statement on the title screen. *"It is our job to search for them."*

I had needed to read it a couple more times before it sunk in. Was this really a genuine organisation dedicated to scouring the world, looking for *monsters*? And by monsters they didn't appear to mean Idi Amin or paedophiles; they meant *monster* monsters, with fangs, and glowing red eyes, and wings and claws, and all that jazz. It took some further research before I convinced myself that it wasn't an elaborate hoax. As I delved deeper into the world of Fortean zoology, something I'd buried many years before, beneath the sediment of adult responsibility and social conditioning, was starting to stir. Something once fuelled by a subscription to *The Unexplained*, and avid viewings of Arthur C. Clarke's *Mysterious World*; a long-dormant, wide-eyed, juvenile optimism that wanted to believe there's *something out there* – something more exciting than the mundane drudgery and diet of *The X Factor* and *Big Brother* and Friday night takeaways that most of us are happy to accept as our lot. And I knew, almost immediately, that I wanted to be a part of it.

An exchange of emails led to me being invited down to Devon to meet with the CFZ team. Over tea and tequila (mercifully, not in the same cup) we talked cryptids, big cat encounters, and expeditions, and before I knew what was happening I'd somehow been pressured into drinking a revolting, fruit-flavoured liqueur, and secured a place on a planned trip to Guyana to search for giant anacondas, a bipedal apeman known as the 'di-di', and something called the water tiger.

As the CFZ's preparations fell into place, over the following months a surreal quirk of serendipity unfolded. For those who don't share Jon's enthusiasm for my questionable oeuvre, I used to review video games for a living, and still write a monthly column for gaming bible *Edge*. Capcom, meanwhile, is one of the world's biggest games companies, and in the summer of 2007 it was due to release a new game in its *Monster Hunter* series. Suddenly, somehow, the CFZ had a new sponsor, and the Guyana trip was a go. And I needed to get my bloated backside down to the gym, and look up Guyana in an atlas. Like everyone else I've spoken to since, I thought Guyana was in Africa.

"Where is it you're going?" they would ask.

"Guyana," I would reply.

"Oh, in Africa."

"No, it's in South America."

"I could've sworn Ghana was in Africa."

"It is, but I'm not going there. I'm going to Guyana."

"Where's that?"

"South America. I just told you. Guyana's on the north-east coast, squished in between Venezuela, Brazil, and Suriname."

"I thought Suriname was in Africa. Or India. Or somewhere."

"Shut up now please."

Outside of the CFZ, the only person who seemed to have even heard of Guyana was the manager of my local comic shop. His father was born and raised in Guyana, but *"He got out the first second he could, because of the crime"*. This was not reassuring. With no dedicated travel guide freely available – frustratingly, Bradt had one scheduled for release two weeks after we were due to return – I had to dig deep to find what I could about the country. Firstly, I was surprised to discover that Guyana was the inspiration for Arthur Conan Doyle's 1912 novel *'The Lost World'*. The flat-topped Mount Roraima, Guyana's highest point, which sprawls across its border with Venezuela and Brazil, was the mysterious, cloud-wreathed plateau Conan Doyle imagined playing home to dinosaurs and cavemen. Though he had never visited Guyana himself, the writer had met with the explorer Colonel Percy Fawcett, who later disappeared while fannying around in Brazil, looking for lost cities. Conan Doyle had also read of the exploits of the awkwardly-named Sir Evarard Ferdinand im Thurn, Governor of Fiji (and later Guyana itself), and the first man to lead a successful expedition to the top of Roraima.

Less happily, I learned, Guyana was also the site of the Jonestown Massacre. Perhaps mixing it up with the Waco siege, for some reason I'd always assumed that this incident had occurred on US soil. In reality, in 1974, mentalist preacher and US tax exile Jim Jones had taken himself, and almost 1,000 of his Peoples Temple of California followers, down to Guyana, where he founded a community in the middle of the rainforest. Four years later, 909 of them – including 287 children – were poisoned, in an apparent case of Kool Aid-assisted mass-suicide. The town still stands in the North of Guyana, albeit all but swallowed by the jungle. I was sorry to learn that it is becoming something of a grisly tourist attraction.

More or less the only other things I learned are that there are a lot of waterfalls in Guyana, and that it is home to pretty much the largest species of every animal guaranteed to provoke a girlish reaction in, well, idiots like me. Guyana's goliath bird-eating spider is the biggest spider on the planet, the largest specimen of which had a 28cm leg span (perhaps it did only eat birds, but I didn't much fancy waking up to find one squatting above my face). The titan beetle can grow up to 15cm long, and has a pair of jaws powerful enough to snap a pencil, and slice through a human finger. Then there's the world's largest butterfly (wing span 21cm; I didn't fancy *that* flapping around inside my tent in the dead of night), the world's largest leech, and - yes - the world's largest snake. On top of this it boasts some of the world's poisonous snakes - including the 14ft long bushmaster, whose Latin name, '*Lachesis muta muta*' means "brings silent death", and the riverbank-dwelling fer-de-lance, responsible for more human deaths than any other South American reptile. Then there are the black caimans, (the largest predator in the entire Amazon basin), piranhas, vampire bats, and several species of very deadly scorpion. And that was all before you factored in the possibility, however remote, of encountering a hitherto undiscovered species of ape-man, and some sort of aquatic big cat thing.

Nevertheless, the biggest threats were from the smallest things: the mosquitoes, and sand flies, who carry a parasitic disease which causes flesh to rot, and your extremities to drop off. It was little wonder why the Guyanese interior was barely inhabited. Half of Guyana's 8,000 species are found nowhere else in the world, and new ones are being discovered all the time. You could see why it inspired Arthur Conan Doyle, and I fully understood why the Centre for Fortean Zoology decided it might be the perfect place to find a "monster".

"*I prefer the waterproof tent with insect screen mesh with sleeping mats and sleeping bags system. To me the hammocks are great for areas with no insect pests - namely mosquitoes - but by night vampire bats can still get you (rabies might not be fun).*"

With Richard's blessing, I'd sent an eMail to Damon Corrie, the Arawak Indian tribal chieftain who was to be our guide in Guyana, and this was his response to my concerns about taking tents. My worries had stemmed from a TV documentary, in which Ewan McGregor and tubby survival expert Ray Mears (who, frankly, looks as if the only thing he's ever foraged for is crisps) visited a Central American rainforest, for no discernable purpose. They'd slept in hammocks, but Damon - via Richard - was recommending tents. As the weeks went on, and the Guyana trip loomed closer, I was becoming increasingly nervous about my preparation, or lack of it. Mental alarm bells were ringing intermittently, and I wanted to know first-hand whether tents really *were* the best option, because even if you take away the insects, the amoebic cysts, the caiman, and the ape monsters, I remained anxious enough about the camping aspect alone.

As far as such behaviour goes, camping is an alien world to me; one populated by boy scouts, and bearded, greying men in loosely knitted jumpers - often in the same tent. I've always considered the thought of visiting the Glastonbury Festival too extreme. The first of only two times I've gone camping was on a trip to the US when I was 13. My parents and I had flown over to visit my sister, who had somehow married an American called Jimbob. He had been raised in the swamps of Georgia (I'm not making this up), wore dungarees, drove a pick-up truck, and thought it sensible to take two shotguns along when we inexplicably decided to spend a week of the three-week trip living under the stars. We camped out in the woods of northern California, where my sister one night left her shoes outside the tent. In the morning, her laces were gone. Whether they had been eaten by wildlife, or stolen by a woodland lunatic, we did not know, but it was my conclusion that if someone, or something, was capable of pilfering laces from a pair of trainers, what was to stop it getting into my tent, and eating half my face off?

The second time I went camping was as an 18-year-old. Full of drunken, teenage bravado, a friend and I had got it into our heads that hitch-hiking around Britain, with a vague plan to visit some standing stones, would be a good idea. We somehow ended up lost on a moor on the Isle of Lewis, in the middle of the night, in the worst storm the Outer Hebrides had seen in years. Genuinely fearing for our lives, we walked until we found a remote garage-cum-breakers yard, climbed over the fence, and sheltered for the night inside the rusting shell of a coach. We woke the next morning to discover that the garage was owned by Popeye's arch-nemesis Bluto, who, as you might expect, didn't take kindly to finding mud-encrusted English teenagers sleeping in his rusting coach shell. And we didn't even have any spinach to hand.

Suffice to say, the two incidents were enough to put me off ever again doing anything as stupid and pointless as sleeping in a tent. If we were meant to camp, God would never have invented hotels. Unfortunately, where we were going, the nearest hotel

would be several hours walk away. Armed with an opening paragraph full of brackets, Damon Corrie didn't hold back in telling me just how basic things were going to be.

"There is no indoor accommodation once we leave the urban centres of Georgetown or Lethem (tents every other place). No guest houses in the areas we are going to. Impress on everyone that fact that once we get into the Pakaraima mountains meals might be prison fare (to you guys). Farine (cassava granules which need no cooking) is the base, taso (salted & smoked) beef might be the main meat we can get to buy in the area for any camp-outs, when we are fortunate enough to overnight in a village (like Taushida for example) we should be lucky enough to have chicken and rice etc, and will endeavour to buy local fruit whenever possible.

No Itinerary really from the time we reach Toka (drop-off point) on day 3 till day 12 when we should get picked up to begin the journey back so as to NOT miss your flight back home. So about 9 days in the middle of nowhere eating whatever we can get and enjoying it. Dangers, theoretically are limitless, but in my 8 years venturing into these areas I have not had any serious mishaps, no broken limbs, no venomous snake bites or scorpion stings, African killer bee swarms attacking me, bandits robbing us and dumping our dead bodies in the river etc, but all these risks are out there. I've only had malaria once, and amoebic dysentery twice.

Work on carrying your pack around for hours on end - up and down stairs might be best for training. If you are a fussy eater - lose that habit real quick, no menu of choices out there. Get insect repellent - but remember you will be applying it to your skin every bloody day all day, so check the internet/your doctor for the cancer causing elements in some brands.

Generally I am in my tent by sunset, no campfire chats unless the location is free of the bloodsucking flies and insects. There are about 4 types of 'vampire' flies, from needle head size to 1cm. These big ones are diurnal but bite a 2mm diameter piece of skin out of you every time and you certainly feel it immediately. The tiny ones have some crap in their saliva that leaves you with a pin-head size black dot stain, and mine have not gone away since I was bitten 3 weeks ago. These little buggers get you morning - noon - and night it seems.

Wildlife - always luck and chance, I have had some very memorable sighting experiences - and other times not seen anything. Definitely will meet local people, mostly from the Macushi tribe, some former rebels who tried to liberate their territory from the Guyana government of the Stalinist dictator Forbes Burnham in 1969 - as this area is a stronghold for retired & closet guerrillas. Bandits are scarce here, they (bandits) are non-Amerindian criminals from the coastal cities who tend to hide out in areas of diamond and gold extraction to prey on the miners. The areas we are going are not big mining areas, mostly small time operations known to a handful of tribesmen.

Slippers or sandals are good for flat walks or in village jaunts, hiking boots/sneakers for the hard going. Get velcro rather than laces, as every time we reach a stream or river we will have to wade through it (not encountered any rivers deeper than waist height (1 metre) in these mountains yet), and wet shoes will really destroy your feet. I walk barefooted if I have to - but I am sure my feet are harder than yours, I use coarse grain sandpaper to get my dead skin off my heels! After 30 miles of the rocky trails even my bare feet begin to bleed though - but that was my first trip in this area when I was in my 20's.

Do like we do, do not let it get you down or in a foul mood, always try to lighten the mood, even if it borders on eccentric behaviour such as telling jokes to yourself or remembering humorous moments in your life. Negative energy saps everyone around, positive energy has the opposite effect. But basically, I am betting money that it will be the hardest trip in all 6 of your lives."

Vampire bats, amoebic dysentery, malaria, guerillas, bandits, rabies, cancer, killer bees, scorpions, starvation, depression, 30-mile walks, monsters *and* eccentric behaviour? By this point I was worried that the trip was fast becoming a really stupid idea, and I was seriously questioning why I'd ever agreed to go along. I was trying hard to put my fears in a lead-lined box, and do my best to prepare. Unfortunately, those preparations weren't going great. My plans to get in shape hit a brick wall when I discovered that my US Army jungle boots (£70 from eBay) had the cushioning properties of an iron maiden. I got two miles into one walk before I could feel the blood sloshing around inside, and had to head back. I was in too much pain to make it all the way home, and quite literally tumbled into my parents' house, where my poor mother was subjected to the full, ghastly state of my feet. I had shredded the skin off both big and little toes, the arch of my foot was in a similarly raw state, and from my mother's horrified reaction I may as well have turned up at the house with a tramp's carcass draped across my shoulders. Not only did this mean investing in a more expensive pair of boots (£150 from an online extreme adventure store), but it put my feet out of action for over a fortnight. Not only could I no longer build up my stamina, but I'd have to delay breaking in the new boots. I'd just got my feet back in working order when some over-zealous toenail-cutting led to my big toe getting so badly infected that I could barely pull on a pair of socks without howling like a punched dog. It required a course of antibiotics, and another fortnight or so of no exercise, to get back to normal.

Additionally, I'm not awfully proud to admit that I'd also hurt my bottom. In a bid to counter years of sitting upon my fetid backside playing video games, and watching DVDs, I'd joined a health club. In the first fortnight of membership, fear of jungle-based collapse foremost in my thoughts, I pushed myself particularly hard on the exercise bike. Unfortunately, the exercise bikes at my local gym were not built for comfort. After a week of such punishment my buttocks were ground to a nub, I had a painful fissure at the base of my spine, and my groin was shredded to the point where wearing underpants was a profoundly uncomfortable experience. Effectively bedridden, I felt helpless, with to do nothing other than surf the web for camping equipment, and further evidence that I was going to die in some country I couldn't even point out on a map. If such punishment is what it took to get fit, I was happy staying a couch potato.

An £80 water purifier was tested on my garden pond. The water came through a suspicious shade of grey, and left a revolting taste in my mouth, which was still there two mornings later. I also purchased a pair of leech-proof socks, which made me look like some sort of mutated, circus clown, and which were only likely to contribute to my inevitable heatstroke. The thought of opening up my boots to find a family of fat, old leeches feasting on my foot, was enough to stall any sense of vanity. Heatstroke or cosmetic humiliation I was prepared to handle. Shrieking like a little girl in front of half a dozen local Amerindian guides, because I'd found a leech sucking on my testicles, I was not (though I eventually left the socks behind).

Vanity aside, in the interests of safety I offered to go halves with the CFZ on the cost of a satellite phone. Though they'd wanted it in order to dispatch daily reports for the expedition blog, which Jon Downes was preparing to update daily back in Devon, I just wanted a lifeline to my family back home, and the knowledge that - if things really did go tits-up - I'd be able to call in a helicopter to get me the hell out of there. The phone didn't come cheap, and contributed to stretching even further my already over-stretched expedition budget, but for me it was the most essential piece of kit. Not only for safety reasons, but as a lifeline with normality. I'd also had my jabs – including an unpleasant reaction to my yellow fever shot, which, as promised by my doctor, had given me a 24-hour bout of aches, pains, high temperatures, and shivering. I had an even *more* unpleasant reaction when I totted up how much this trip was costing me. I hadn't even made it to Guyana yet, and already I was suffering, both physically and financially.

The full, unbridled, account of my Guyana experience has been put aside for another time – and besides, within these pages the day-to-day details will doubtlessly be better addressed, from a fortean angle, by Richard, Jon, Chris and Lisa. Instead, it seems a more timely use of my chapter to share a few specific moments with you, from the point of view of a first-time monster hunter. Weird fox-thing encounter or not, I don't mind admitting that I approached Guyana from the standpoint of borderline sceptic. Before we left Heathrow I didn't really know the four people I'd be travelling with, and I worried that they viewed me with suspicion, or harbour the notion that I was there to snigger at them. I may have known that I wouldn't commit myself to trudging round the arse-end of nowhere if I wasn't prepared to give it my all, but they didn't know that.

By the end of our two weeks in Guyana I think it's safe to say that we had bonded like a proper team. After initially feeling like a fifth wheel, I think I eventually found my role within our group, all of us bonding through a passion for the unknown, and extreme physical duress. With the exception of the seemingly superhuman Dr Chris Clark - who, I'm sure he won't mind me saying, was also feeling the fatigue by the end - and Jon Hare, who somehow made it through in spite of wearing a pair of unsuitable slip-on shoes that he'd borrowed after mashing up his walking boots, we were all suffering at the hands of Guyana's indescribable heat and unforgiving terrain.

Richard had practically had his extremities chewed off by mosquitoes, Lisa had apparently broken her thumb by falling off a mountain, and her feet were consumed with blisters, while I hadn't really recovered from a mighty bout of heatstroke.

Though I'd only had it the one time – poor Richard suffered most days – as a consequence of sitting beneath a savannah sun with only a Bill Bryson book for shade, followed by a five-mile hike, it was one hell of a thing. Richard had avoided it that day by hiding in a swamp, but by the time it came to pack my tent, in preparation for the dusk walk to our next campsite, I was in a seriously bad way. It took me close to half an hour to roll up my sleeping bag, so addled were my faculties. I felt so wretched that night that I didn't even come out of my tent to eat the caiman our guides had caught for dinner earlier that morning. Unfortunately, they insisted I have some for breakfast the next day...

Earlier in the trip I'd also had a half hour moment of homesickness, which Richard and Lisa had talked me down from. We'd just undertaken a gruelling three to four hour walk back from the village of Taushida - where we'd heard tales of club-wielding di-di tribes, grinning, red-faced bush people, and several accounts of water tigers. Atop a nearby mountain, we'd been the first people outside of Taushida to see a burial chamber, discovered by a local boy only a few years before, and we had hiked out to Tebang's Stone, a local sacred site. It had been an incredible few days, but by the time we reached Toka, our stop-off point before our planned trip into the savannah to find anaconda, I felt like I'd seen and done enough. I wanted to sleep in a bed, and eat a decent meal that wasn't covered in flies, and not feel dusty and filthy.

My low ebb didn't last long - it was literally for less than an hour that I wanted to bail - but when I voiced how I was feeling to Richard and Lisa they were so nice to me that I almost blubbed. After experiencing their kindness, there was no way I was going to abandon the expedition, and my companions, mid-way through. And I was very glad I didn't.

Over the next few days I experienced my aforementioned first-ever bout of heatstroke, got within ten feet of a giant anteater (which we later learned could've disembowelled us with its claws), ate a cashew, got a temporary tattoo, and - at last - visited the proper rainforest, where we hunted for a red, two-tailed alligator in a *bona-fide*, Indiana Jones-style cave. They were experiences that I'll take with me to my grave, but there was a specific moment where I finally became a believer. Not so much a believer in monsters - everyone I've met in the CFZ will tell you that they're there to gather evidence before believing wholesale in anything - but a believer in *why* the guys do this. It was while we were in the rainforest, meeting with Ernesto Faris, the fish farmer who told us about the red, two-tailed alligators. For the umpteenth time in days we were speaking with someone who, with seemingly no agenda other than conversation, was telling us of his encounters with the local fortean zoological population: di-di, giant anacondas, and those red-face bush people.

Almost everyone we met had some story about the bush people, but it was the way they told those stories that I found so intriguing. They didn't talk of them in the way that people talk of the Loch Ness Monster, like it was some fanciful local legend. They talked of the bush people as a statement of fact. They just *were*. Kids saw them in bushes every morning on the walk to school. Other people had seen them while fishing. Even Damon Corrie - who hadn't, in the course of the entire trip, expressed an opinion about our quarry one way or another - finally admitted to having seen one on a camping trip into the savannah some years previously. As Ernesto spoke of hearing the bush people walking around at night, while he slept in his hammock, I felt a swelling of excitement in my stomach (it must've been excitement - there was precious little else *left* in my stomach). It was then that I knew why the guys fly around the world, battling with physical duress, to look for things that, frankly, less enlightened people have already decided don't exist, before even giving the notion the time of day. The monster hunters do this because it's fun, and it's exciting, and because nobody else is doing it.

After we flew back to Georgetown it was time for the CFZ guys and I to go our separate ways. Sick of lugging it around with me, I'd given away most of my gear to Damon Corrie, so further camping was out of the question. Plus, I was hungry, I was exhausted, I was filthy, I was weary of having every vehicle we boarded breaking down, and I badly missed my kids (phone calls home were becoming increasingly tough, hearing how upset my youngest daughter was – it was the longest I've ever been away from any of them). We had gathered a huge trough-full of evidence, and the team all concluded that we'd achieved everything that we could in Guyana (at least until we all return during the rainy season…) I managed to convince the other guys that we could try and bring our connecting flight forward, and possibly treat ourselves to a night or two in Barbados. Frankly, we'd earned it. Unfortunately, as it turned out, there was only one seat on the plane, and so I had to endure a couple of harrowing days lazing on the beach in a Caribbean paradise, while the others lived it up with the tarantulas and scorpions in Damon Corrie's ancestral village of Pakuri.

However, there was one more night in Guyana for me. Once the others bundled into a pick-up for the two-hour drive to the reservation, I was to spend the night in Georgetown with Marvin, a cab driver friend of Damon, and an urbanised Arawak. Of all the Amerindians we'd met in Guyana, Marvin was the one who had most firmly embraced the culture of the West. He'd lived in Georgetown for ten years, and it was only a couple of times a year that he visited the village where he grew up. Despite its reputation for crime - confirmed by everything I'd seen of it thus far - Marvin loved living in Georgetown. His home - while only about as cosy as my dad's garage - boasted a huge stereo, cable TV, and a Nintendo console. His kids were excited about Christmas, flicking through the toy section of what passes for the Guyanese equivalent of the Argos catalogue (presumably the toy section comes after the 'semi-automatic handgun' section). In Marvin's eyes, why would you want to live in a thatched hut, when you could live in the relative luxury of a single story, three-room house made out of breezeblocks? Despite the open sewer behind the house, the lack of glass in the windows, the outside shower and toilet cubicle, and the fact there were more insects in the bedroom than I'd seen during my entire time in the savannah, it was the closest I'd felt to home in a fortnight (even if I did end up making the grand mistake of agreeing to accompany Marvin on his nightly cab rounds through the hellish environs of Georgetown – give me a horde of marauding hominids over the borderline riot of the city's centre any day).

For dinner, Marvin, his wife, and I drove to a nearby fish n' chip restaurant (part Harry Ramsdens, part Oktoberfest, part civil war). Given that he had seemingly turned his back on much of his heritage, it came as little surprise that when I asked Marvin about the di-di he laughed it off as nothing more than "superstition and legend". Expecting much the same response, I mentioned - with a matily derisive snort - the red-faced bush people.

"*No, they're real,*" said Marvin, matter-of-factly, before popping a plantain chip in his mouth. Cue those hairs on the back of my neck, and cue them again when I returned home, and decided to read up on the history of Guyana. I picked up a copy of Sir Walter Raleigh's travelogue, '*The Discoverie of the Large Rich and Bewtiful Empyre of Guiana*', written in the late 1500s (yeah, mate - you may have been a great explorer, but you couldn't spell for toffee).

In a text littered with derogatory comments about the hated Spanish - basically, they're a bunch of drunkards and rapists - Raleigh, who was mightily impressed with the country, wrote: *"Whatever prince shall possess it, that prince shall be lord of more gold, and of more cities and people than either the King of Spain or the great Turk"*.

Unfortunately, no endless reserves of gold have ever been discovered in Guyana, and Raleigh's conviction was seemingly based entirely upon him finding a rusty trowel on a riverbank, which the explorer decided was the sort of thing somebody might own if they were in the business of refining gold. However, Raleigh did encounter first-hand many Amerindian people, and they told him of another race of people, one more elusive, who Raleigh – perhaps fortunately – didn't meet with. He wrote of this strange tribe:

"There are a nation of people whose heads appear not above their shoulders; which though it may be thought a mere fable, yet for mine own part I am resolved it is true, because every child in the provinces of Aromaia and Canuri affirm the same. They are called Ewaipanoma; they are reported to have their eyes in their shoulders, and their mouths in the middle of their breasts, and that a long train of hair groweth backward between their shoulders. The son of Topiawari, which I brought with me into England, told me that they were the most mighty men of all the land, and use bows, arrows, and clubs thrice as big as any of Guiana."

Hairy, club-wielding, mighty men? Sounds familiar, Walt.

Think what you will of those who have dedicated their lives to hunting monsters, but if nothing else during my inaugural experiences with the Centre for Fortean Zoology I've learned to always keep an open mind. Ten years ago I saw something on the Isle of Wight that I couldn't explain, and maybe it *was* just a fox. But the fact is, I sincerely hope it wasn't. I hope that I did see something unexplained on that hillside, something – dare I say it – *fortean*. Why? Because to embrace the alternative is to shut off the possibility that the world is far less strange and exciting than it should be. And that would be a load of bloody rubbish.

CHRIS CLARK

In November 2007 the CFZ team, drawn by stories of giant anacondas, decided to go to Guyana to investigate. The snakes had been seen at a place far in the interior called Corona Falls; the estimated length of 40ft would have made them as long as the creature in the notorious *Anaconda* film, though presumably they were less aggressive and did not have that snake's ability to eat three cast members a day. As well as the core team of myself, Richard Freeman and Jon Hare, we were joined by Lisa Dowley from the Gambia trip, and a new member: Paul Rose ('Mr. Biffo') a writer and internet journalist who had become fascinated by the CFZ, and had resolved to join the next trip.

We had to change planes at New York. My passport contains two Afghan and two Pakistani visas, plus one each from Iran, Syria and Libya; it looks as if I had signed up for the `Axis of Evil World Tour`, and I had hoped that as transit passengers we would be exempt from the usual formalities. I was wrong: we had to fill out two separate forms, containing so many warnings about false declarations that I felt that in order to enter the spirit of the thing properly I should sign them in my own blood. This gave us the right to queue up in order to pass through flight security once again, with the usual routine of removing our shoes and inspecting our toothpaste. Only then could we carry on to Georgetown, the capital of Guyana.

At the airport we were met by Damon Corrie. The arrangements for the trip had been made through his travel company, which specialises in off-the-beaten-track tours in Guyana and Barbados. On the Corrie side, according to a family tree on the wall of his home, he is a descendant of the same King Duncan of Scotland who was murdered in his sleep by Macbeth. Through his uncle on his mother's side he is Paramount Chief of the Eagle clan of the Sarawak Indians of Guyana. Now based in Barbados where his parents live, he runs a number of businesses, is President of the Caribbean Herpetological (reptiles and amphibians) Society, and speaks vigorously for the Guyanan Amerindians in their frequent disputes with central government. His help and the hospitality of family members made our trip both possible and enjoyable.

He took us to his home village with the charming name of St Cuthberts Mission, but known locally as Pakuri after a species of tree. In this pleasant little place, where we could swim in the wine-red waters of a local creek, we heard the first of several stories about the di-di. This curious creature was one of the secondary aims of the expedition. Although descriptions often vary, it is commonly described as being like a large hairy man. Apparently one had abducted a child only two years ago at a place 30 miles away.

We went back to Georgetown to catch a bus to Lethem, not far from the Brazilian border. The bus trip to Lethem should have taken 12 hours: it actually took nearly 16. Fortunately, I spent the entire time asleep, and remember almost nothing about it. At some point in the night the bus broke down and after a long delay we transferred to another, but although I must (presumably) have been awake to do this it has been entirely wiped from my mind. From there we soon went on by truck to a small village to collect our guides, porters and a contingent of teenage girls to do the washing and cooking.

We left the village, carrying our packs, for the long walk into the interior and our encounter with the world's largest snakes. As the day grew hotter and the climb through open country without shade became steeper, some of the party began to have difficulties. Mr Biffo, who had not been on one of our expeditions before, began to appreciate some of the demands made on the practicing cryptozoologist. Richard, more seriously, suffered from heat exhaustion, turning bright red and nearly passing out several times. All this time the teenage girls ambled up the slope swinging their handbags. When we reached the high plateau, we stopped for the night at a tiny village and put up our tents. At this point we realised that we had gone only six miles in the day. Our destination, Corona Falls, was over 60 miles away, and it was plainly impossible to get there on foot in any reasonable time, while the river was currently too low to reach it by boat. The only remaining hope was to rent a helicopter, or aircraft, when we returned to Lethem. We washed in a small stream nearby, being occasionally bitten by tiny fish; that night we saw a bush fire on the other side of the stream.

About five years ago a hunter discovered an Indian grave on top of a nearby hill, and we decided to go and see it. The hill was very steep: near the top Lisa slipped over on a rough slope and landed so hard that she believed she had broken her thumb. The grave itself was at the summit, under a natural slab of rock. The earthenware pots in the grave contained the skull of a boy, and some of the bones of an adult man. The date of the grave is unknown: we took away a broken piece of pottery to be brought back to England for thermoluminescence dating.

Here, as well as other accounts of di-di sightings, we heard the first stories of a race of pygmies with red faces. Kenard, one of our guides, claimed that one had lived in the area as recently as the 1970s, and that he would catch a lift on the back of his uncle's motorbike! Damon had also heard of these creatures: naked and brown-skinned, they had a fondness for tobacco.

We visited a local feature known as Tebang's Rock, a huge upright stone that looked very dramatic in the twilight; Tebang himself was a malevolent spirit, with a habit of infecting children, who would sit on the rock playing a flute made from their bones. After this we set off in earnest in search of anaconda; by this time we would have been satisfied with mere 30-footers. Circling a swamp, we found large furrows where they had passed through, but no sign of the creatures themselves. Kenard always carried a bow and arrows: after going off to look for game, he came back carrying a dead cayman, about two or three feet long, that he had killed instantly with a shot through the back of the skull. Cayman is perfectly edible, and we grilled and ate it that night.

Around sunset one evening we were lucky enough to see a giant anteater on the pampas. The guides headed it towards us, so we were able to see that it had a baby on its back. They have strong arms and powerful claws to enable them to break into termite mounds. Damon told us about a man who had come across one and, presumably aiming for a Darwin Award for anti-survival behaviour, tackled it with a club: the creature reared up and crushed his chest, killing him outright and proving that you don't need teeth to be dangerous.

By this time the sun was so hot, and the lack of shade so unpleasant, that Jon and I set off to walk to a local ranch a few miles away. It was worth the walk: we got a cold drink, and sat in the shade watching people sawing steel reinforcing rods into short lengths by hand, while back on the grassland the others were standing in the swamp in order to get under the only shade. Conditions were generally difficult on this trip: the heat of the day, which made trekking difficult, dissipated only slowly at night, making sleeping in the tents uncomfortable. Leaving the flaps open caused an invasion of small biting insects that seemed to have a particular delight in Richard's legs. Cassava, an 'edible' root which formed the basis of our diet, is dull in all the various forms in which it is presented, and we could drink only from water purifiers.

We returned next day to the point where we were to be taken back to Lethem. While we waited, Richard heard an interesting first-hand account of the 'water-tiger', an unknown possible felid and another secondary aim of the expedition: the witness described it as being different from the giant otter, which he had also seen. At Lethem itself we decided to fly back to Georgetown rather than attempt the bus again; while buying our tickets we learnt that the only helicopter in the country was not available, and we had no time to charter one from Brazil. After a night in a local hotel, we drove out into the jungle and stopped at a clearing. We were met by a man called Ernest who ran a small fish farm, and knew a good deal about the creatures of Guyana. He had seen a 30ft anaconda only ten years before; he had had a friend who had seen the di-di (and blamed it for his death); and many years before he had himself seen one of the red-faced pygmies (which he believed to be natural colouration rather than paint). He also said that a few years ago he had seen a cayman smaller than any known variety. Since the place was nearby we set off to explore. It turned to be a very large pile of boulders, containing a small cave with openings at the front and top. We climbed in and examined it, but it showed no sign of ever having been occupied by anything.

The next day we caught the plane back to Georgetown, staying in a very pleasant small hotel while we waited for the flight home. This gave us a chance to look around the capital. Guyana is one of the most racially mixed nations in the world, with populations from Europe, Africa, India and China as well as the Amerindians. Perhaps for this reason, it seems to have no distinctive character of its own. Certainly, apart from the very striking Anglican cathedral built entirely from wood, Georgetown is not a place you would go out of your way to see. Particularly discouraging was the bar we passed with a sign outside displaying a long list of banned weapons. We visited the local Anthropology Museum. The museum contained a number of Indian artifacts including a large bark canoe, and an interesting display, which disputed the standard account of the migration to North America having come solely across the Bering Straits in favour of multiple entry points. Damon Corrie had told us that while termite mounds were being demolished to make way for the new airport, some had proved to contain bodies, as though they had been buried there. The curator knew nothing of this: when he was told that Damon was the source, he dismissed him abruptly as unreliable, and Richard and Lisa had a characteristically vigorous disagreement with him.

We pressed on to the Natural History museum, where Richard rebuked the staff over the conservation of their exhibits. At Georgetown Zoo a third dispute, this time over the treatment of their animals, was only avoided because no responsible person was in evidence. The anacondas, an aquatic species, were being kept in a cage almost entirely dry. Unhappier was the lioness enclosed in a space inadequate for an animal half the size and in obvious distress. Some of the smaller animals such as the ocelots were more comfortable, and we were able to get much closer to them than a safety-obsessed Western zoo would have permitted.

Our flight home avoided New York and went from Barbados, where Damon Corrie's parents live and where the better infrastructure allows him to conduct business more readily. Barbados is ranked first in the developing world by the United Nations: certainly I have rarely been in a more attractive place. It has good climate, great beaches, friendly people, a general relaxed atmosphere and the impression that here things actually work; the colonial legacy of good government has not been frittered away in corruption and instability. We were given a free tour of the island, and then taken back to dine with Damon's family before going to the airport.

Was the trip a success? The primary aim, searching for giant snakes, was not even attempted: demanding conditions and inadequate preparation made this impossible. However, we collected a number of accounts of other recognised cryptids, such as the di-di and the water tiger, and even of quite unknown creatures such as the small red-faced men and the tiny cayman: perhaps we have broadened the field of cryptozoological knowledge even if we failed to deepen it.

THE CENTRE FOR FORTEAN ZOOLOGY
GUYANA 2007

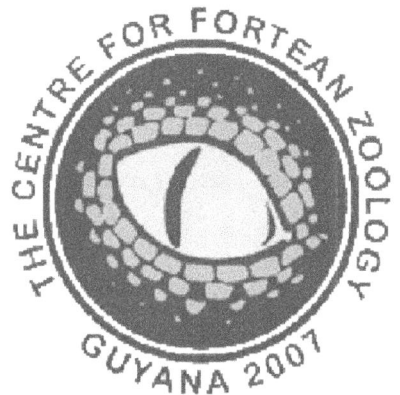

JON HARE

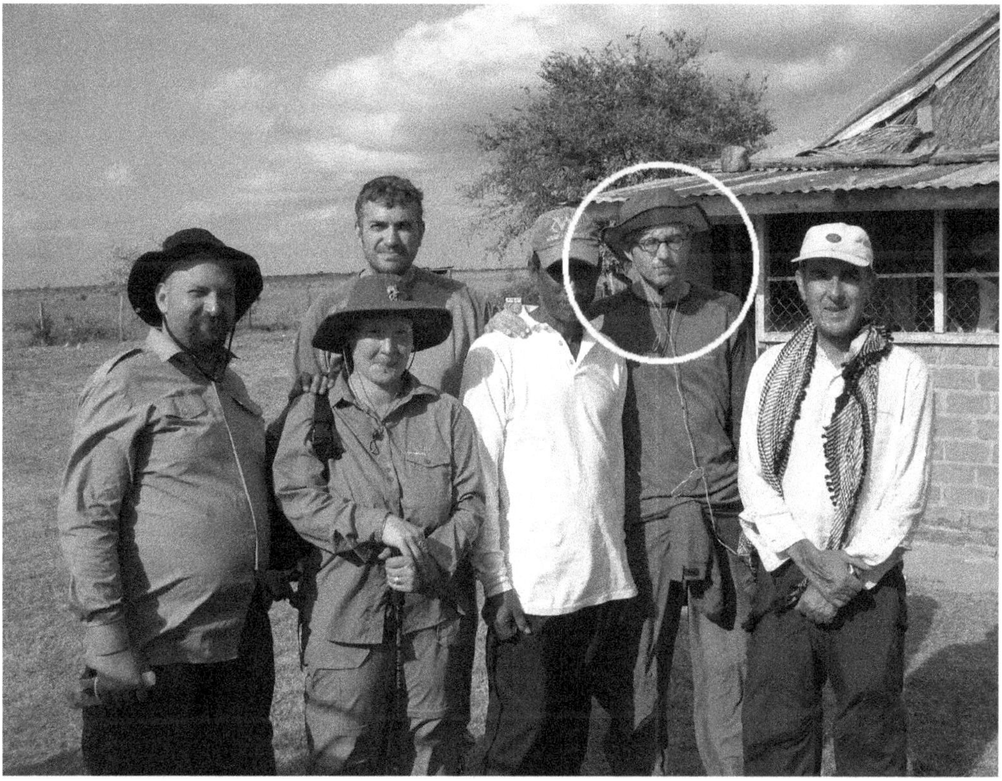

Alright, so we're bouncing in the back of a knackered truck in the Guyanese highlands, or rather, in the tracks around the base of the hills that constitute that middle-ground between the lowland forests and the mountains - flat plains covered with burned prickly scrub and the grey rotten teeth of termite mounds - and David, our indecently boyish half-Indian guide, points to one of the mounds and casually lets slip that many of the mounds have been found to contain skeletons according to some long-forgotten ancient Arawak burial practice. And as we cling to the slippery sides of the vehicle and try not to fall screaming into the road to be flayed by pebbles and left for the crows, all I can think is:

People eat dirt
Insects eat dirt
People are eaten by dirt
People are eaten by insects
People are eaten by dirt eaten by insects

People do eat dirt, but not just any dirt. Not all dirt is the same. White clay - rich in calcium, minerals and nutrients – and sulphur and phosphorus - calming to the stomach - is sought out and eaten by pregnant women in East and West Africa, and the African American populations of the Caribbean, in Asia and East Asia, and in Central and South America. Do that regularly enough in the West and you'll be classed as a sufferer of pica - a psychiatric disorder marked by the compulsive ingestion of non-food items. Lightbulbs, nails, hair, insects, and octopi. However, in Africa, you can buy chunks of food-clay on market stalls and in local pharmacies. For pregnant women, soft clays sooth the stomach trouble of morning sickness, and the calcium provides nutritional supplementation for the two following trimesters (*when the skeleton of the foetus is forming*).

Clay builds skeletons. Skeletons are buried in clay. Clay destroys skeletons.

While they'll put you away for gobbling earth, many modern medicines for stomach upsets contain elements derived from dirt. Bentonite clays, kaolins, and attapulgites. The clay coats the gastrointestinal tract and can absorb toxins, as well as helping to prevent nausea.

I once saw a television programme about a woman who consistently gave birth to babies with exciting degrees of Autistic Spectrum Disorder. Four of them, as I recall: tow-haired, intelligent, hyperactive little maniacs, all pre-teen and existing in an unpredictable behavioural oscillation, from angel to demon in a single nightmarish second. The mother's control of them was largely dietary. Stick the wrong thing in their mouths (a soft drink containing some chemical or other, or a boiled sweet full of additives, or an apple with the wrong shiny spray-on glaze) and you might as well have fed a mogwai after midnight.

Years of bruises, breakages and bitter experience created the perfect diet, and all was calm. But, like Jurassic Park, LIFE FOUND A WAY. Suddenly - inexplicably - maniacal hyperactivity returned, with wallpaper being peeled off, and glasses shattered with the screams, while children tore through the air like fleshy cabers lined with razor-sharp milk teeth. But the diet was unchanged, so what had happened? On investigation, it turned out that the children, deprived of the substances necessary for proper crazybastard functioning, had been secretly eating the inner lining of their nursery curtains. A material which, a few 'phone calls confirmed, was laced with the same chemicals responsible for their original mania.

They just knew. The children had an inexplicable and uncontrollable craving to chew on the curtains, the same way a pregnant woman can't explain her vanilla-icecream-on-rye-with-brown-sauce sandwich. Cravings aren't conscious, but the body knows what it needs.

So, as I was saying, people sometimes eat dirt.

Termites eat dirt too. Pulped and swallowed and churned in the intense alkaline environment of the insects' five-chambered stomachs, the substance is retextured, smoothed, and changed. The regurgitated soils inside the termite mound are bonded together with organic cement to create natural concrete. Corrosive acidic clays surrounding the mound, that could dissolve flesh and bone, are mollified as they are modified, calmed by the termite's stomach factory and expelled, creating a haven - an environment of neutral, or mildly alkaline soils. The earth in the mounds contains a clay content some 20% higher than the surrounding area with higher levels of extractable bases, organic carbon and mineral nitrogen. Plant matter in the soil is digested, purified and enriched.

Of all the various traded clays on the African market stalls, none is so sought after as the dirt from inside termite mounds. Eating dirt was so widespread in Africa that plantation owners in the Caribbean, observing the practice among their slaves, sought to stamp it out by fitting their property with elaborate mouth locks, like the opposite of a scold's bridle, keeping crap from flying in rather than flying out. Slaves were labelled Clay Eaters, Dirt Eaters, as were the poor whites of the southern states of America who adopted the practice.

Magic termite earth is used in African kilns, the chemical properties of the dirt making it resistant to the very high temperatures necessary for smelting glass beads. The sacred fires of Australian aboriginal peoples are made from sticks, grasses and crumbled termite mounds. Why?

More than just a producer of strange soils or digestive remedies, the mounds themselves have an unparalleled permanence in the landscape, a ruggedness and imperviousness to the environment unmatched until the invention of concrete in the late Roman Empire. Raised slowly, imperceptibly, they last generations, timeless as the hills, slow as geology.

Soldier ants build mounds out of their dead. The ant colony's dead-or-dying percentage, a substantial fraction of the colony, is scooped up by workers, and the dead bodies gathered together into neat piles by the side of the column - hills of motionless chitin. People, too, raise mounds to honour their dead. The American continent is covered with barrows and artificial hills; temple mountains on which dead bodies are exposed to be picked clean by carrion birds, their flesh scattered to the four winds - sacred stones covering the bones of the honoured dead.

And so, after exposure to the elements and the scouring away of all meat, organs and connective tissues, it should come as no surprise that peoples such as the Australian aborigines, the Warlpiri people living to the North and West of Alice Springs in Australia's Northern Territory, will hack their way into the heart of their flat and rectangular, bizarrely magnetic termite mounds, which resemble fields of large, reddish gravestones, and place the skeleton inside.

When the runway was extended at Georgetown airport in Guyana, a number of the country's lowland termite mounds, grey and tall like melted limestone stalagmite, were demolished. The local Indians manning the bulldozers were unsurprised, David said, when skeletal human remains were revealed inside many of the mounds. And while local archeologists scoff at the idea, and from what we were told, the practice seems to have died out among the local Indians, its memory is preserved in local tradition and in the skeletons crouched, encysted, within the mounds.

The acidic clays of local red soil will etch bones to nothing in a single generation, unlike the benign alkaline environment of the termite mound, and this is what I mean when I say that people are eaten by dirt eaten by insects, and it should be no surprise that people might inter the bodies of their dead in these transformative, magical, solemn and indestructible termite mounds. The dead man's existing house, a fragile affair, will be burned to the ground and scattered. His new dwelling, as eternal a resting place as man and termite can devise, will stand forever in the hard shadows of the outback, in the shade-less Guyanese savanna, and in the legends of local peoples.

"My friends are shit, that's about the size of it"

THE CENTRE FOR FORTEAN ZOOLOGY
GUYANA 2007

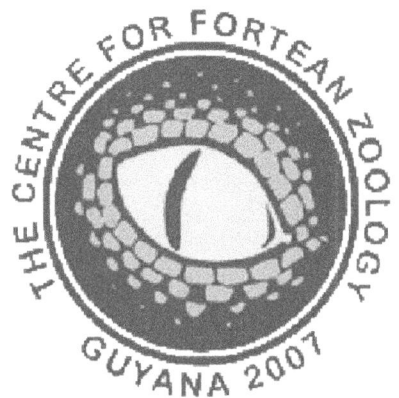

LISA DOWLEY

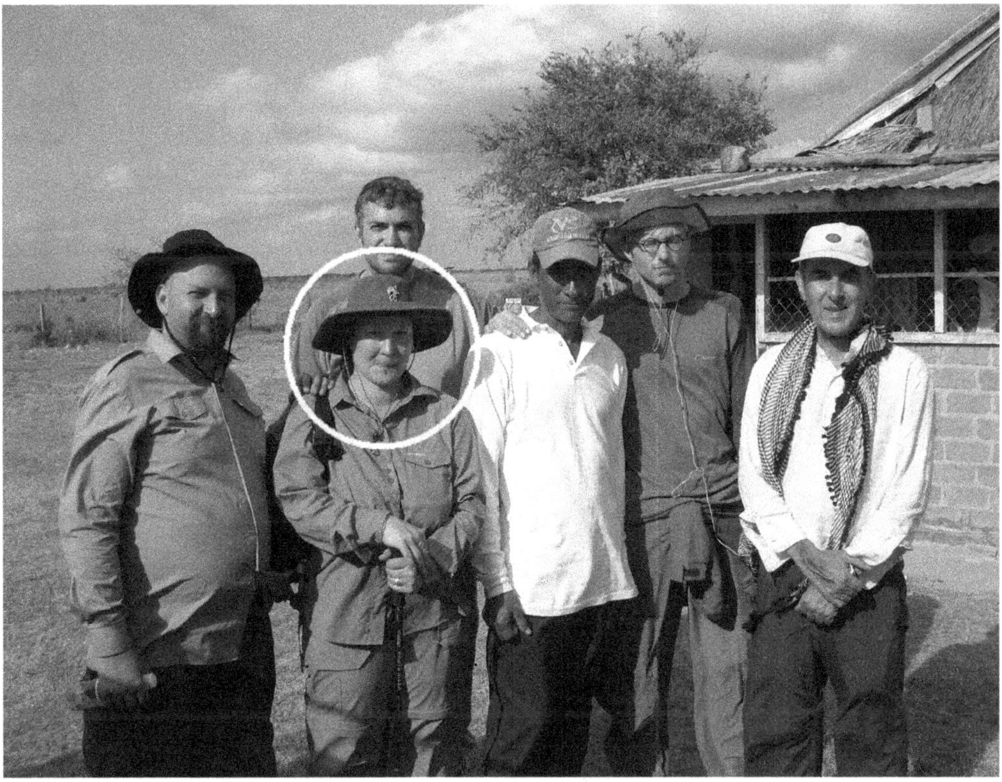

Well it seems unbelievable that a year has passed by so quickly, and that I find myself again preparing for another expedition with the CFZ. However there's something different about this one.

For the first time in the CFZ's history we have secured a sponsor - a very respectable one too - via Sam Brace (thank you Sam) who at the time worked for the software giant Capcom (if you don't know who they are, they are the guys responsible for successful computer games such as *Street Fighter, Resident Evil* and *Devil may Cry*).

What's even better is that there are no strings attached to this funding; no clauses stating that we have to take along cumbersome film crews, or that we have to sign away our images and rights. Just *"there's the money chaps, go find some monsters or whatever it is you do, oh and good luck!"*

The only condition of the sponsorship was that the expedition was scheduled to coincide with the release of their latest game *Monster Hunter 2*. Not too much to ask for by anybody's standards, and you must agree that it is a bloody big achievement. The CFZ, which, granted, is the largest organisation of its kind, is a wholly voluntary, self-funding establishment, attracting huge corporate sponsorship such as this - a bit of a coup if you ask me.

The expedition team this year comprised of Richard Freeman (crypto-zoologist and writer) Dr Chris Clarke (astrophysicist) Jon Hare (science writer and self-styled musician) Lisa Dowley (that's me - adventurer, amateur archaeologist, photographer and filmier, domestic goddess... oh the list could go on, but don't worry it will not - I'm sure you get the idea that I will turn my hand and have a try at anything) and a newcomer to the team, successful TV writer and author Paul Rose (Mr Biffo).

Paul's current publication, I am assured, is a most amusing book titled, *Confessions of a Chat Room Freak*. Paul has been, and still is, involved in many aspects of the media from writing scripts for *Eastenders*, to being implicated with supplying additional material for Lenny Henry's latest show, but my favourite of his creations was a BBCtv pilot show called *Biffovision*. If you were a bemused child of the 70s as I was, its homage to classic Saturday morning television highly entertaining. The series was never made, so watch it on YouTube.

Richard and Jon (Big Boss man of the CFZ) - both long time fans of his work - came across Paul's witty writing skills via *Digitiser*, which was a UK Channel 4 teletext magazine which should have been about video games. Uninterested in the computer game aspect of the programme, Paul seized the opportunity and used it as a platform for his humour in facilitating his own brand of obtuse jokes.

This was to be the expedition team, quite an eclectic mix by anybody's standards, but this *is* the CFZ. What did you expect?

As the title of the book implies the expedition was to Guyana, but to search for what?

You may well ask.

Over the last 18 months or so, new and recent reports have be coming in to the CFZ office of outsized anacondas. This was further spurned-on by Richard learning of a company that specialised in guided cultural tours which offered excursions to known, and remotely explored, areas of the interior of Guyana where such creatures have been spotted. *Guided Cultural Tours* are operated by a chap by the name of Damon Corrie.

Richard made contact with him, and began correspondence regarding research and background information. It turned out that this man was nothing less than a hereditary Amerindian Chief of the Arawak Clan (one of the nine Amerindian tribes of South America). He is also an active pro-Amerindian rights campaigner who has, and does, travel all over world championing their indigenous rights and fighting to have the Amerindian voice heard. He has fought long and hard in a bid to get the Amerindians recognised as the true indigenous people of Guyana, hence to say he is not overly liked by the Guyanese Government, but - rather - tolerated. He is a much respected and admired figurehead of the native people. As if none of that is impressive enough, he is a committed self-styled conservationist who specialises and breeds Guyanese reptiles and invertebrates.

Via this correspondence, Richard was told by Damon that, as recently as last year (2007), outsized anacondas had been seen at a remote large body of water known as Corona Falls. Damon had spoken directly with the Amerindian hunting party that had witnessed the large serpentine creature. They were so fearful (the anaconda considered by some Amerindians as an evil spirit - an equivalent to the Devil) of its immense size that they had fled the area. While quizzing one of the members of the hunting party regarding the size of this snake,, he replied that a mature dead palm tree of some 30ft was floating in the waters, and he could see the head and tail, which stretched beyond either end of this floating tree, which - at a guess - would make the said snake an incredible sum total of 40ft!

So it transpired that the headliner was to gather as much information as possible: witness statements, and actual footage - if we could - of outsized anacondas.

During the weeks that led up to the expedition, Damon and Rich exchanged a great deal of information, and mention was made by Damon of a number of other cryptids, such as the di-di. This is said to be a large hairy bipedal creature that bares similarities to that of the sasquatch and yeti. It had been seen in various locations in the interior, and Damon knew of a few Amerindians that had had encounters with these large hairy bipedal man-beasts.

He also imparted knowledge on a little known creature that had been labelled the water-tiger. This is a fearsome creature that strikes dread into the very hearts of the Amerindians, and is said to live in, and around, the many murky brown water courses that traverse the mountainous interior.

The CFZ expedition itinerary seemed to be shaping up rather well. This was the first time the organisation had been to South America, and we were going to utilise the opportunity to the max. Before long, we had a full list of cryptids to investigate, and - sure enough - before too long, the day of departure arrived.

The flight was at 6.15pm from Heathrow on the 14th November 2007. As it was an evening flight, it gave us all plenty of time to arrive in good stead. I was the first to arrive, due to the fact that I live the furthest away; in Staffordshire. I caught my National Express Coach, which - thankfully - was taking me directly to Heathrow. Great, I thought. No long protracted journey full of stops at grotty remote bus stations. I alighted at Heathrow around 10.30am, and was greeted by a whole host of redevelopment work that the airport was in the process of. I made my way through the temporary white hardboard corridor that had been erected in a bid to blank out the construction (and, I would like to think, overweight and ill-fittingly clothed workmen) in progress. However, it did nothing for the sound of prolific drilling and banging that resonated up and down the corridor. I made my way to the *Virgin* stand in order to get my bearings, and familiarise myself with the surroundings. Once that was done, I wandered off in search of a reasonably priced outlet to grab a late breakfast.

Time passed soon enough, and I made my way back to the *Virgin* check-in stand to meet Richard, who was followed a few hours later by `Mr Biffo`. Then along came Jon, and lastly, but by no means least, Dr Chris. We all decided to pass the time away by having a meal and drinks, and it wasn't long before the conversation turned to our impending excursion: our hopes, thoughts and what we may find.

We boarded the flight on time. This was the first time that I have travelled with *Virgin*, and I must say I was pleasantly pleased. The facilities they have are really quite agreeable, and once I had got the hang of how the in-flight entertainment system worked, I settled down to watch my favourite comedy of the moment; *The Mighty Boosh*. I drifted off to sleep with thoughts of explorations to the Arctic, accompanied by polar bears, little people, and a softy spoken southerner with androgynous dress sense.

The journey to Guyana was pretty much uneventful, apart from some confusion over form-filling at New York airport. As we were in transit we were under the impression that we did not have to fill these forms in. However, it soon became clear that we did, and this would have been a rather straightforward exercise if it had not been for the fact that the correct forms could not be located straightaway. Anyway, they were located, filled-in, and duly handed in along with passports. We then went on our merry way. It was quite late at New York Airport, and most stores were in the process of closing down, so I grabbed a coffee, and we made our way to catch our connecting flight to Georgetown, which is the capital of Guyana.

Now, I think at this juncture the pronunciation of Guyana should be clarified. Guyana is pronounced as it is spelled, *Guy*-ana and not *ghee*-ana. It is an Amerindian word which means "land of many waters". As the plane flew over in preparation to land at Georgetown, its apt name was more than evident, for out of the 'plane window could be seen a profusion of rivers, creeks, and other bodies of water. They seemed to hold the lush chenille-like deep green jungle together, as if it were nature's very own patchwork quilt; the waterways - the stitching holding it altogether.

This land-mass is located in the northern part of the Amazon Basin of South America, so it's no surprise that there should be many such waterways to the Atlantic Ocean. There are three main rivers that run through Guyana: the Essequibo, Berbice, and the Demerara. It is on the eastern banks of the estuary of the Demerara River that the capital and chief port of Georgetown is located; this site was chosen originally by the Dutch for strategic reasons.

Guyana, formerly British Guiana, is the only English-speaking country in South America. Its neighbours are Venezuela to the west, Brazil to the south, and Suriname (formerly Dutch Guiana) to the east. Although on the South American continent, Guyana is regarded as the only country in the West Indies - the rest of the West Indies are island states.

At 83,000 sq.miles (215,000 sq.km), Guyana is about the size of the United Kingdom. Of this area, some 90% of the population live in the coastal belt, some 270 miles long by 10 to 40 miles deep. This area is below sea level at high tide, so to stave off the relentless onslaught of the Atlantic Ocean, sea walls and sluices, or kokers as they are known, line the coastline. About 70% of the country is still tropical rainforest, and this geographical area is located south of the coastline, and above sea level. Savannah country is located to the south and south-east of the country. The latter two geographical zones of Guyana are especially rich in minerals (not to mention timber) including bauxite (used in the production of aluminium), gold, diamonds, kaolin and manganese, to name but a few. This richness of the land is literally proving to be its undoing as vast swathes of land are deforested and exploited by, in the most part, large American and Canadian corporations. Even the coastal region does not escape from the intensity of cultivated rice and sugar cane production.

With its location being just north of the Equator, it's in the tropical zone of the Northern Hemisphere. As a result, there are only two season types – dry and, yes you've guessed it, rainy, with the latter occurring twice a year in May/June and December/January. However, due to global warming, these seasons in recent years have become somewhat disrupted and blurred to say the least, and nobody seems to know what consequences this is having on the flora and fauna. Could it be the reason why there has been a sudden upsurge in sighting of various cryptids? From the research and background reading that I had undertaken, most report that with the commencement of the dry season, larger anacondas (quite possibly females as they grow larger than males) were thought to move to remote pools which do not dry up, hidden deep within the jungle's interior. They are stimulated into movement to larger bodies of open water by the seasonal rains. For a great many creatures, the seasons also regulate mating times - if the seasons become out of kilter, what's the effect on the fauna? These thoughts - and many others - ran through my mind as we touched down in the early morning sunshine of Georgetown Airport.

Somehow we managed to find ourselves at the back of the queue, though none of us were quite sure why it was not moving. We seemed to be there for hours, but in time our passports were checked and we made our way through to the front of the airport. I spotted Damon; sitting at the far end of the entrance lounge a tall slender man wearing three quarter shorts, a white vest t-shirt, and a small woolly Benny-style hat. We made our way over and exchanged greetings. Damon then escorted us out to the minibus he had waiting for us. We all hopped on and were then driven through Georgetown, which I must say at this point looked nothing more than an upmarket shanty town, and was less than inviting. It seemed to have the appearance of being worn out and of having definitely seen better days and times. The supporting background of the radio in the minibus did not fill me with great confidence either. An advert for a concert was broadcast; at the end it said in a very casual way *"Remember no guns please"*. We all looked at each other and burst into fits of laughter - it kind of summed up Georgetown really. After the journey through the main town, we then changed the minibus for a pickup truck at a local stop, and made our way to the reservation. It was great to sit on the back of the pickup and get fresh air blowing all around us. The journey was long and bumpy, and I found myself at various points clinging on to our backpacks so that they did not fall off the pickup.

We arrived at the reservation late morning. Its original name was Pakuri (derived from the trees that were once plentiful in the area). In recent times it had been renamed, St. Cuthbert's Mission, due to the catholic missionaries' involvement with the reservation; however Damon preferred it to be referred to by its original title, and later that day took great pride in showing us the last remaining pakuri tree which stood solitary – alone, but proud. Very much like the Amerindians, I thought. The forests in this region have been heavily exploited, especially for wallaba and other commercial timbers. Pakuri is located 60 miles up the Mahaica River. This area is part of the savannah belt that stretches out behind the alluvial coastal plain. It is easily accessible by a 4X4 vehicle, except in the rainy season, when the whole area floods.

The route to the centre of the village was a twisting, turning, bumpy, and - not to say - uncomfortable affair through the natural bleached-like soft white sands of the creek, which at times blinded me with its stark brightness. In between the sands were pools of muddy sandpits, as Pakuri's location is within a creek, which is subject to regular floods dependant on the amount of rain. On route, we passed local Amerindian dwellings, which were houses situated on stilts in a bid to deal with the regular flooding, and to cut down on the amount of encounters with neo-tropical rattlesnakes in the home, which is a common occurrence here.

Part of Amerindian culture is to bathe twice a day, and Damon expected us to respect this, and do the same. After all the travelling, I was more than eager to get into some cool waters. He led us up the dirt track, which led to the main body of the creek - which never dries up - which was used for bathing. And at the far end this was separated by a small sand bank and reeds - the washing of clothes took place here.

The water was of a deep reddish brown due to the tannins (leaf/plant debris etc). It was wholly refreshing once in the cool reddy brown waters, and it was rather amusing to see Jon's amazingly pale skin gain a deep red hue, while submerged in the cool waters. While in the waters, Damon told us that in the past caiman and anacondas had been seen in, and around, this particular part of the creek. We made jokes about this, and the candiru fish, which I may add was not present in this stretch of water.

After bathing, we were escorted to the main hall of the village to sign the visitor's book, and then back to a house that was owned by some of Damon's extended family. It is here that he keeps various creatures that he has found, and that the locals bring to him.

Damon showed us a rainbow boa that he had captured just under two weeks before our arrival. He said it was quite vicious when he first had it, but in such a short space of time it had become quite used to being handled, and appeared to enjoy it. We all took turns holding it, or maybe it took turns in holding on to us, as its grip was amazing and proved quite difficult at times to release. It was a rusty colour phase that I had not seen before, with only a slight hint of the iridescence it was renowned for.

We were then privileged to be shown a scorpion, which Damon had found. It was thought to be a completely new species, which had a vague green hue to it. It was so new that it had not been officially described, nor had it been given a scientific name. What made it all the more fascinating was that her back was full of tiny scorpion babies! I took the opportunity to take as many photos as possible.

While sitting in the shade of the palm trees we were told, via Damon, a number of di-di, water tiger and anaconda stories, and about the people who had encountered them. Approximately two years ago, in an Amerindian village some thirty miles from Pakuri, two children - a brother and sister - were walking home from school. The girl was around 12 and the boy a little younger (school finishes shortly before midday) and they were walking along a path in some scrub savannah. What the boy then described as a 'huge hairy man' stepped out of a crop of trees and grabbed the girl, turning and quickly making off back into the denser wooded area. The girl was never seen again. There was no subsequent police investigation; it was treated as 'just one of those things' by the authorities!

Another story was told whereby a man from Pakuri had seen a di-di, but only its back view, as it walked away from him some seven years ago. He was not in the village, so we were unable to interview him directly.

The di-di and the water tiger, though cryptids, are not totally unknown. Accounts of these beasts have been documented from the 19th Century up to the first half of the 20th Century.

An interesting chap by the name of Matthew French Young, (1905-1996) chronicled his life and times in the Guyanese interior. For over 50 years this English educated man worked in the wild-forested interior and savannahs of Guyana as a diamond prospector, gold-panner surveyor of the uncharted bush, hunter and builder of roads for himself and large corporations. He first became aware of the water tiger via nearly getting himself arrested. He had elected to go for a swim in the dark muddy brown waters near a place called Tukeit. Being young and fearless, he dived into the waters and swam out to some rocks in the middle of the river. He was recalled by loud shrill blasts from a police whistle, and frantic gestures from his travelling companions. He was then warned never to swim in the dark brown waters again as there were things under the water dangerous to anything moving, such as the water tiger, which was known locally as the 'Massacuruman'. It was said that its shape in the water bore a resemblance to that of a human. It was reputedly covered with hair, having small ears set in a head, which was armed with the fangs of a tiger. Its feet and hands were webbed, and set with horrifying claws; it also had a long tail. In total it was said to be some 5ft in length. On another excursion into the interior he was shown the handy-work of said beast. At Chinapow Landing he was shown an Amerindian woman whose face was hideously disfigured. As a small child she had be playing and bathing at the river's edge, while her mother was tending to the washing of clothes. All of a sudden there was a terrifying scream, and hearing this, the mother turned to see a 'Massacuruman' come up out of the river and attempt to grab the young girl. The mother rushed over and managed to save her child from death. However the claws from this mysterious creature's hands left horrific wounds, scarring the young child's face for life, and maybe to serve as a warning of what lies beneath the dark brown muddy waters. I wondered if we were likely to encounter such tales.

Damon's brother-in-law Foster, who was also to be one of our chief guides while in the interior, told us that several years ago - in a watercourse a few miles from the village - he saw the trail of a huge anaconda. He said: *"Judging by its width it would have been far larger than the 20ft stuffed specimen in the National Museum in Georgetown"*. However, this does not take into account that the creature may have recently eaten and was looking for a place to rest up, and let the digestive process take its course.

It was now late afternoon, and we made our way back through the upmarket shanty that is Georgetown, through the dingy soiled back streets that just oozed with seediness and violence, until we found the pick-up point for the overnight bus to Lethem. This pick-up point was, to put it politely, horrid and grubby. It came as no surprise to me that the bus, when it arrived, was suitably matched to its environment - dirty, rusty and grubby. *"How far do you think we will get before it breaks down?"* I exclaimed.

The reply from the guys was *"not very bloody far!"*

I boarded the coach, passing by the seats that had the environmentally friendly air conditioning (broken windows) and elected to sit further to the back of the bus. I sat down, and literally sank into the seat - there was no padding or springs - and only the Gods could have possibly known what was growing and living in the worn-out draylon that covered the seat. To be honest, I was not over bothered with the state of the coach, as the journey was beginning to take its toll, and I felt extremely tired.

Sure enough, no sooner had I dropped off to sleep, I was awoken by a bizarre sound, and yes you guessed it, the bus had broken down. It seemed that the coach could not be repaired so we all waited some four and a half hours for the replacement coach to arrive. I did not venture off the coach, but opted to stay in my seat, and drift in and out of light sleep. In time the replacement coach arrived, which I may add was a great improvement on the previous one. The seats had padding; they were covered in a type of plastic, which cut down on the thoughts of them having their own eco-system within them. I boarded the coach, found a suitable seat, and settled off to sleep again.

I awoke at dawn, and witnessed the terrain change from deep lush jungle, to more open areas of tropical forest where logging was taking place. I was suitably shocked and horrified to see beautiful mature huge trees - some in excess of 15ft in diameter and at least 12ft high - cut down and left on their sides awaiting removal. It struck me, as we passed, that the other trees seemed to be lamenting their demise as they swayed in the gentle early morning breeze.

The landscape soon changed, however, to more open grasslands and savannah. We came to a ferry crossing that was to take us across the Essequibo River - the biggest river in Guyana, *"at some wider points this river"*, Damon told us, *"has islands in it that are larger than Barbados"*.

I did wonder how long we would be here for, as I saw the driver take a fishing rod from behind his driving seat, and go and sit on the riverbank and begin to fish. However, the ferry soon turned up and towed us across in a very sedate and calm manner. On the other side, we all took our places on the coach, and continued on our dusty, bumpy journey to Lethem.

After some thirty plus hours of travelling, we had eventually arrived in Lethem. Getting off the coach, I exclaimed *"Is there any chance we could fly back to Georgetown when it is time to go back?"* Dr Chris heard me, and said: *"I totally bloody agree with you there"*. I must say that the thought of a comparable return journey did not fill me with enthusiasm. Damon said he would make enquiries.

It had been organised that we were to spend a night at a guesthouse, before starting the expedition proper with a two-hour drive to the village of Toka the next day. Damon and Foster were staying with relations in the Amerindian settlement that was on the outskirts of Lethem town. That afternoon I made full use of the shower and bed, as it would be some two weeks before we would have such comforts again. Later that evening, after we had all refreshed ourselves, we walked into the main part of the town and found an internet café, and duly sent messages to the CFZ that we had arrived at our start destination.

The next morning after breakfast, we all assembled and waited for the pickup truck to take us to Toka. We waited and waited. It was a little late in turning up to say the least. At this point, I realised that we would be hiking in the heat of the day, which I had wanted to avoid at all costs. In time the open-backed truck arrived, and we began our two-hour drive to Toka. The landscape *en-route* was amazing; it was so flat, with mountains in the distance on both sides of the road. Now and again you would get an outburst of green in the form of bushes or trees. In the distance, just off the main track, bizarre shaped termite mounds - for all the world looking like mud encrusted Christmas trees - could be seen. Their size ranged from small lumps to mounds that were in excess of 6ft tall. We saw a variety of birds along the route, jabirus, caracara; I even spied a macaw, as it flew high above us, then veered off into the distance.

We reached the small village of Toka around ten o'clock. My earlier fears were confirmed that we would be walking through the heat of the day with little, or no shade, a full backpack (mine weighed in excess of twenty four kilos) plus all the water I could carry. We waited again, for some time, for more guides and domestic help to join us, which pushed the time on even further. At last we made a start along the thin trail which petered out into open rocky mountainous terrain, which bore a striking resemblance to images of Mars – dry, dusty and red, not to mention the rocky landscape with little or no shade. Our destination was a remote village called Taushida. By the time we had got into the swing of the walk it was high noon, and the heat was unrelenting and unbearable for some of the group.

There was no shade to be had apart from the odd bush come tree. I found I could cope with the heat as long as I kept moving, it was only when you stopped that the intensity and ferociousness of the sun became apparent. Poor Richard, it proved too much for him and he suffered terribly, passing out on at least three occasions. It was only six miles to Taushida, but those six miles took us nearly five and a half hours to complete due to all the stops, and the fact that the group were at various stages of the route. The time of day we were walking did not help matters, and all this combined took its toll. I was one of the first to arrive at a small river just outside the village. I did not have to be asked to bathe, I was more than eager to get in and cool off, my legs ached, not from the walk or the heat, or even the weight of my pack, but due to how uneven and ever-changing the terrain had been. Soon we were altogether again, and the cool water soon lifted our spirits. We had fun with the little fishes, cichlids I think they are called, as they cheekily nipped at our toes. After a most enjoyable toe dip we loaded ourselves up again to make our way on the last half-mile to the village. Foster stopped me and said: *"I will carry your pack; you have done well to walk this trail"*. I was rather taken aback by this statement. It turns out the first time Foster had attempted this trail he too had suffered badly, and it seems I had impressed him with my sheer determination. I was somewhat hesitant to give him my pack to carry, I felt guilty, but he insisted. May I add at this point that Foster was carrying his own gear, the camp's gear *and* two backpacks! These Amerindians are most undeniably a breed apart - their endurance and stamina knows no bounds.

As I clambered up the last incline, I heard a bizarre crackling/hissing sound to my right, and at first I thought it may have been a snake. A bit late in the day I thought, but as I turned to look down what I was actually witnessing was the start of a natural bush fire! I thought how amazing - the rawness of nature and how intense this heat must really be. I quickly then moved on, not wanting to be 'roast Lisa' so early on in the expedition.

We reached the village of Taushida around 2.30 in the afternoon, and we pitched our tents and made camp around a tree, which was the centre of this part of the village. The actual village was separated by a stream, which cut deep into the rocky dry ground. On the small, inhabited bits of land on various sides of this water course, houses could be seen; very different from the ones of the Pakuri settlement. They were built at ground level from mud bricks in the most part. We were now in Makushi territory - another one of the Amerindian tribes of South America. In times gone by the Makushi and Arawak tribes formed a great fighting allegiance and this deep friendship remains to this day.

After we had pitched our tents, we were invited to bathe properly down at the creek, and this we all took great delight in. We spent the best part of what was left of the afternoon in the cool waters of the creek. I also took time to get into the habit of regularly topping up my supplies of fresh water via the purifier I had brought. The small creek was very picturesque, but apart from the tiny cichlids, there were no other fish to be seen. I enquired about this when Damon joined us, and he said that further upstream they had been over using haiari vine to fish, but due to its constant use it had killed off all the fish. This haiari vine is collected by men and women, then cut into 2ft lengths, placed in bundles, then carried to a flowing stream. The said bundles are then placed on the rocks and pounded with a heavy stick. The juice from these bundles of vines runs into the water turning it a milky white colour. This milky substance drugs the fish and causes them to float to the surface where they are then collected.

On returning from our daily ablutions, what seemed like a sumptuous feast had been laid on for us: mild curried beef, chicken, cassava and rice. All in all it was a most enjoyable feast; washed down with of all things, tropical flavoured 'Cool Aid'. During dinner, we were told that Corona Falls was a full 70 miles away. The reality of this would mean we would have to cover at least 20 miles per day (each way) which would mean walking through the relentless heat of the day. There was no way Richard could make that distance in a couple of days, six miles had almost killed him! I doubt if any of us could, if the truth be known. The conversation turned to the possiblity of renting a helicopter once back in Lethem. We decided to spend some extra time in Taushida, as the area was ripe for exploring.

As soon as the sun sets it's dark. It's a darkness rich in beauty as it is void of any light pollution, and standing out against the darkness, my attention was drawn to a bush fire burning on the distant horizon - it seemed to be blazing on the trail we had walked some hours before. We all watched it for some time, mesmerised by its dancing-like movement across the darkened landscape. We were then told by Damon that while he was here a few weeks ago, checking out the area, he had been witness to what he described as 'balls of light', in the sky above the mountains. At first it resembled a bright star then seemingly shattered into several smaller balls of light. He said they remained visible for some time dancing erratically in the night's clear sky. He then went on to say that over the other side of the valley, above the opposite mountain range a thunderstorm was in full throw. Were the two connected? I pondered over this question while looking at the clear night's sky. All of a sudden a meteor/shooting star flitted across the sky, burning away into a nonentity. On that note I chose to retire for the evening, I clambered into my tent, and settled in for my first night's sleep of the expedition.

The next morning I was the first up (in a bid to get some private bathing time) just before daybreak, which had its bonus as I was able to see some bats (I think they were the infamous vampire bats, but could not be totally sure). They fluttered around my head, and the only tree in our camp, before returning to their mountain caves, no doubt after a night's hunting. After breakfast, we set off in the early morning coolness to seek out a burial cave which had been discovered by Moses, a young Makushi, while out on a hunting trip.

The route to the cave was pleasant at first, but the further we went and the higher we got, the faint trail petered out and we found ourselves clambering over, and under, sharp thorny bushes, scaling the ever rockier and sparse grassy landscape. We stopped half way up the mountain range to admire the view, and - to be fair - it was a damn impressive view of the valley below.

But, as I neared the top with the cave mouth in sight, following in a young Amerindian's lighter and swifter footsteps, the long grass I stepped on gave way to.... well nothing, and my foot disappeared by about a foot and a half, between two rocks. I felt a sharp pain in the sole of my left foot as if impaled on something, then due to being off-balance, I began to tip over to the right, and started tumbling off the mountainside in an ever rapid motion. I cannot recall exactly what happened next, but remember falling, bashing my shoulder very hard, and crushing my right thumb in the process.

I grabbed at what I could to stop the fall, which happened to be the long mocking grasses which covered the mountainside, and which had caused my fall in the first place. Again they had the last laugh as, as I grabbed at them and clung for dear life, my hands slid down the long grassy trails, cutting my wrists and the palms of my hands to bits, but I *did* manage to stop falling any further. In total I must have fallen around 17ft. I tried to make my way back up the mountainside, but again my left foot had become entangled in the long grasses, and wedged again amongst some rocks. All the others - by this time - were inside and exploring the burial cave, and were blissfully unaware of where I was, or indeed, what had just happened to me. Fortunately, Dr Chris was at the back of the mountain trail, and spied me over the mountain edge and clambered down some way and gave me a hand getting back up. I got to the mouth of the cave, sat down, and gathered my thoughts. I think I was in a little shock as the full gravity of the situation dawned on me as I looked out over the view of the valley. As I sat there nursing my black cherry-coloured, ever-swelling, broken thumb, two beautiful scarlet macaws flew past. The sight of these magnificent parrots helped me to compose myself. I did my best to put the pain of my injuries to the back of my mind.

The climb (and the fall) was well worth it. It was not a true cave, but more of a rock shelter. At the entrance to the cave there appeared a mass of stones that had fallen in, suggesting that at some stage the rock shelter entrance had been sealed up. Inside, towards the back, were two huge pottery urn-like vessels of some two and a half foot in diameter. It was clear to me that these had been previously disturbed. In each of these urns were the remains of one human. Both urns had a similar urn placed on top as a sort of lid which appeared to have been sealed in the past with some sort of clay mixture, which was a very light yellow in colour. In the urn nearest the cave entrance there appeared the remains of a full grown male - the skull and parts of the spine were visible. Damon, and the other Amerindians, were of the view that this urn contained the remains of a possible chief, shaman or holy man, and as such requested that it was not touched, but respected and left alone. Which we duly did.

The second urn was far more interesting, it was somewhat larger than the first one, and had the remains of a young boy (possibly between the ages of 9-12) contained within. Its male status was confirmed as, while Damon and myself looked through the urn remains, I found a peccary tooth, (wild pig) with a hole pierced through it. This would have been a trophy from his first hunt, and mixed in with the burnt debris, was a large amount of very small beads which appeared to be glass - these would have been strung along with the peccary tooth and proudly worn around the boy's neck.

As mentioned before, both the bodies had been extensively burned which suggested a cremation, however the bodies had been cremated at a different location to that of the cave as there was no evidence whatsoever of burning within the rock shelter. So, therefore, the bodies had been transported to this resting place possibly indicating that these individuals were of some potential importance to merit such a gruelling journey to such a prominent, remote position. Combine this with the commanding view from the rock shelter's mouth, which encompassed the entire valley below, and some considerable distance into the mountainous horizon. It also points to whoever buried them had a complex burial ritual and an organised society, which took great concern in caring and burying their dead, as there also seemed to be small vessels which contained the remnants of food placed in there, no doubt for the individuals' time or journey into the afterlife.

The pottery urns themselves had each been placed on a ring of stones, raising them up off the ground, again indicating that somebody had gone to great lengths to ensure that the final resting place would be fitting, appropriate and secure. I also took time to scour the cave walls, roof and entrance just in case there were any petroglyphs but I could not see any.

Damon was not overly sure how old these remains could possibly be, or indeed whose they were, or what tribe they belonged to. They could have been around five hundred years old, or as recent as the end of the Amerindian wars, which ended a little over a hundred years ago. While sifting through the young boy's remains I was allowed to take some photographs with a Bic Biro used as scale to show how small the boy's skull was, (the biro measured just over 5½ in long with the top on). I was also allowed to take some pottery fragments and beads for potential further analysis, but no bones. Because we had taken something from the site, out of respect we had to put something back, and a necklace from one of Amerindian group was placed in a small pot (which was also found inside the urn), along with all the smaller bones then placed inside the urn.

A custom to show respect for the dead here was to smoke a cigar, which none of us in the group possessed so the next best thing - a cigarette was duly passed around, this was rather amusing as the only person in our group that actually smoked was Dr Chris. We all took turns in having a little drag on the nicotine stick; it reminded me of my miss-spent younger days behind the P.E hut where we would gather for a crafty smoke at lunch times.

My views on the burial? It seemed that it was of Makushi origin, as the style of the urn, and actual burial, was very reminiscent of what they had ascribed to in the past. For me, it was doubtful that they were of Carib or Atorid origin as this was not in keeping with their customs or beliefs, as both of these tribes prefer to bury cremation remains in sealed pottery vessels, then bury them in the ground. As for the age of the site, well seeing as the beads had gone through a heat process (cremation) I would say they are glass which could push the date to anything from the fifteen hundreds (they were used for trade by the Dutch) onwards, which was when the Makushi first started living here in the 'Taushida' mountain ranges. Another possibility is that they pre-date the Makushi settlement of the area and that they are of unknown tribal origin. Conversely, they could be as recent as being placed there within the last hundred years. An extensive analysis of the beads and pottery would need to be undertaken to establish a more accurate date.

As we made ready for the return journey, Moses told us that the di-di had been seen walking across this very mountain range some ten years previously. During the walk back, to blank out the pain, I recalled Matthew French Young's near miss encounter with the di-di. While hunting with a couple of Amerindians near Mount Kowatipu, they heard some creature approaching; however this creature's approach was accompanied by jabbering in a deep guttural tone. French Young, and the small hunting party, hid in nearby bushes unaware of what to expect. He then went on to say,

"I could hardly believe my eyes when I saw two huge beasts resembling apes or gorillas. They were covered with a sort of brown hair walking upright like a man. They were about 6 feet tall. When they had passed I noted the size of their footprints with splayed toes.

After they had passed the Amerindians would not hunt in that area again saying it was 'Bushmen' we had seen."

At this point in time I was in that much pain I would have been quite happy for a di-di to pick me up and carry me off to anywhere, as long as I did not have to amble along myself. The walk back was long and painful for me as my injuries became more evident, but I still managed to get back way before Richard, who again had suffered badly at the hands of the merciless heat of the late morning sun.

While waiting for him to return, I got chatting to another guide who had joined the group; Kenard Davis of the Makushi tribe. He had many anecdotes regarding various cryptids known and unknown. I got him to relay his stories again (and captured them on film) once Richard returned from the mountain range. He told us a story that his father had related to him. It happened in the 1950s.

A man had been hunting, and was coming home over the mountains. He had chosen the mountain pass route as it was quicker than walking all the way around the mountain. He was carrying two wild fowl, the spoils from a day's hunting. As he neared the top of the mountain, he looked up and saw a huge hairy man asleep in the trees. The vines seem to him like they had been woven into a hammock-like shape. On seeing this sight, the man became so frightened, that he ran all the way to the bottom of the mountains - *still* clutching the birds.

When he returned to his village he became ill, and belived that the di-di had put a spell on him. He consulted a shaman who, while in a trance-like state, managed to contact the di-di. The creature relayed to him that the man had frightened himself into sickness. It also made mention that he (the di-di) lived on the mountain, along with his wife and daughter who lived on neighbouring mountains, and that they did not wish to harm people.

Kenard stated that he had never seen a di-di for himself, but he did tell us of an unusual creature he had encounterd. Up until the 1970s a tiny, red faced small pigmy type individual was well known in the area of Toka.

He was hairless, in the most part naked, had brown skin and was approximately 3 to 3½ft tall. He had what appeared to be a bright red painted face, and seemed to always be presenting a huge odd grin. Apparently he would leap out of bushes grining at passers by, thus scaring them, though he never did physically harm anyone.

Kenard's uncle had a motorbike and the little red faced man would often jump on to the back and catch a ride. He always jumped off at the same spot which Kenard's uncle assumed was his home. Damon stated that people left gifts of tobacco out for him.

Later that evening, as it became cooler, we were invited by Kenard to visit a sacred site, 'Tebang's Rock'. This was a 30ft upright pillar of rock that stood on the open savannah. Kenard relayed to us the story of Tebang. He was a little man who walked around at night placing his hands upon children, which in turn would make them very ill. Once the child had died and succumbed to Tebang, he would then set about making a flute from their remains and play tunes on their bones while sitting on, or around, the said rock.

It was said that he could be seen on a moonlit night, whistling, shrieking and dancing. Richard seemed to think it had many parallels with an African goblin known by the name of Tokoloshe - a horrid and evil creature with an outsized head that wandered through the night transmitting illness to children via placing his hands on them.

Strangely, while Kenard told the story of Tebang, I took some photos of him and when I looked at the photos more closely, Kenard seemd to be covered in a fog-like mist. However, the very next picture was perfectly ok!

By the time we had reached Tebang's rock and looked around it, the light had all but faded, and we had to make our way back to the camp in near total darkness. All we had were the torches we had taken with us, and it was slighty un-nerving and exciting at the same time, as in the natural blackness of night, we had to re-cross various streams and transverse over very uneven terrain. This played hell with my foot injuries, not to mention that most creatures come out to hunt at night, namely neo-tropical rattlesnakes and fer-de-lances (another type of venomous viper).

We eventually got back to camp and were greeted with another fantastic feast of local food. We dined that night (our last night at Taushida village) on chicken, rice, fish, fresh pineapple and cassava. Cassava is a main staple in the Amerindain diet, and it is a major source of carbohydrates. However, it is toxic unless it goes through a certain treatment process to remove the toxins.

It is shredded, then soaked in running water, and squeezed through a wickerwork tube in order to remove the toxic element. Then it can be dried, and pounded into a granulated form. It is a remarkable substance - a small handful can keep you going all day. After its treatment it can be eaten in a number of forms. That night we had it in a form that was very reminiscent of cous-cous, along with the less palatable cassava bread, which I can only describe as a type of crispbread which has the texture of hardboard. It is also popular as breakfast in muesli-like form, covered in sugar water.

Over dinner, Damon told us about various mysterious locations in the area. He said that he had been told by one man while climbing a mountain that he was almost sucked into a cave very near the summit by what he believed to be a dragon. He did not see the creature, but a great vacuum-like sucking force came from the cave's mouth, almost dragging him in. Damon (and myself) seemed to think that this great force may well have been the wind blowing through the mountain, which is quite likely if it was hollow, or if there was a network of caves located within.

After dinner, I retired to bed early as I wanted an early start to packing as my injuries, I felt, were possibly hindering me in keeping pace with the rest of the group, and the last thing I wanted was for me to be responsible for holding the group up. While in the quiet of my tent, the extent of my injuries became all too apparent to me. I lay there helpless, accepting wave after wave of pain washing over me, starting at my foot, calling in at my thumb, with a pit-stop at my shoulder, then back to the foot to start over again. I fell asleep dreaming about an oddly patterned rattlesnake just staring at me.

I woke early in order to bathe and pack. Richard seemed to take forever to pack. But eventually we set off back to the village of Toka at around 8 am. While walking back, Damon asked us if we had any dreams while staying there. Richard relayed a bizarre dream where he broke into his nasty aunt's house and urinated down her stairs! I told him of my strangely patterned snake that had paid my dream sleep a visit. He then made mention that a rattlesnake had been killed the previous night in another part of the village! Were the two connected? I pondered this thought as we walked on in the early morning sunshine.

There was very little breeze this particular morning, and what there was, blew warmly over us. The pain I was in became more and more intense - step, pain, step, pain, step. I think I had covered about three miles (half the distance) when I exclaimed that I just could not go any further with the backpack on as it felt as if it was physically cutting into and burning. I felt terrible, that I was letting the group down with my ailments. Mr Biffo, bless him, seemed really concerned regarding my condition, and suggested that I should get medically checked out via a doctor. For me that was not an option that would have meant missing out on the expedition, and I really did not want to do that. Dr Chris then offered me some very strong pain relief medication (opiate based) that his doctor had prescribed for emergencies. As a rule I will not touch any form of medication, but on this instance I made an exception, and doubled up on the dose in order to get a kick start into the system as it were. One of the guides, Kenard, also offered to take my pack, to which I reluctantly agreed.

While walking at this slightly slower pace, Kenard relayed many stories about the Makushi and Caribs, and how important the area was with its lookout points, and the many battles that had taken place along this route. He told us that many stone axes and pottery shards could still be found in the area. If I had not been suffering so badly from my injuries, I would have relished the chance of exploring the local area, with the chance of finding genuine stone axes, and original Amerindian pottery.

We all made it back to the village of Toka around 10.30, a total of around two and half hours, which was a great improvement (injuries and all) on our first attempt at this route. We got ourselves into the shade and settled down to rest up, which I was more than happy to do, as I had not rested properly since falling off the mountain.

The bus to pick us up was not scheduled until late in the afternoon. As we rested in the shade, Foster brought me a couple of stone axes which were apparently being used to prop open a door! They were beautiful. One was a hand axe, you could tell just by holding it. You could feel the contours where previous hands had held it and used it, the gentle contours on the stone felt so smooth. The second one was more of a standard stone axe, and you could see where it would have been attached to a wooden handle. They really were works of art and beauty.

Later, we were led to a small creek where we could take our daily bathe. En route Damon pointed out a cashew tree, and he picked a large yellow plump fruit from one of the long hanging branches. Most of the group had not seen a cashew in its entirety; the actual nut sits above attached to the fruit. However, in its natural state, the nut is toxic and needs to be thoroughly washed and wiped before being consumed. The fleshy fruit is a different matter - that can be eaten straight away. Richard bit into it, sending spurts of juice in every direction. Then I tried it. The fruit was rather deceiving, it may have been physically juicy, but on the palate, it was very dry to the taste - very much like cranberries. This dry aftertaste lingered for some time, and I had to wash it away with some water.

In due course our mode of transport arrived - an open back truck - which took us some distance down the forever straight dusty road, to our next destination which was a place called 'Crane Pond'. This was located within the lands of 'Rodrigus Ranch'. These open savannah plains were dry in the most part, but deep within the savannah heartland there still remained open bodies of water, disguised beneath thick layers of water hyacinth. In the rainy season, the whole area becomes one giant lake. It was here, about a quarter of a mile away from Crane Pond, where we planned to camp and go in search of anacondas. Not outsized ones, but we lived in the hope of locating and obtaining some footage of an average-sized one. While walking to the area in the late cool afternoon breeze, we happened across a female giant anteater with a young baby riding proudly on her back. The creature, on catching our scent, made off in the other direction, but Damon raced over, and herded her back in our direction. This was a brave manoeuvre on Damon's part, as these creatures have been known to disembowel humans with their formidable claws. Mr Biffo and myself froze in position with our cameras ready, as we realised that she was heading straight in our direction, but as she neared us she must have sensed us, and moved slightly to the right. She trotted past me at least no more that 4ft away. It was a spectacular sight to behold, I did not realise just how big these creature actually are, and to see the baby on her back was just incredible.

We then moved on to set up our camp and tents as it was beginning to get dark. After we had set up camp, Kenard began to tell Richard and myself stories that he had heard surrounding Crane Pond. It was once thought of as a dragon lair by the local Amerindian cattle ranchers. They used to say, *'don't sleep too deeply at Crane Pond or the dragon will take you'*. Back in the 1950s, when cattle ranching was big business for Guyana, a band of Amerindian cowboys had camped very near to Crane Pond for the night. During the night they heard some huge creature stir and rise from the water. From out of the darkness they could hear its hissing-like breathing, and on hearing this they panicked. A number of them mounted their horses and quickly galloped off into the night. Those who remained, fired their guns pointlessly into the darkness in a vain hope of dissuading whatever was out there from coming closer, giving them time to make a quick getaway. Was this creature an anaconda? They are known to make a peculiar sound when breathing. Anacondas are seldom found far from water, and occupy a very large variety of aquatic environments including rivers, large and small streams, lakes, ponds, swamps, ditches, temporary pools, and flooded forests. In addition, they can be found in areas, such as dry forests, and the llanos, which is a seasonally inundated tropical savannah such as the area of Crane Pond, where we had camped for the night, chiefly because of the abundance of surface water in these areas for six to eight months during the wet season.

Opinions on the size attained by anacondas have been offered by many experts in their field, and explorers, some of who have actually had first hand experience of them. They have been recorded in size from 10.5ft to about 25ft. How large these snakes get is an issue that has been clouded by authors, suggesting that the outsized super-snakes (snakes reported to be in the 50 to 100ft range) are species not currently known to science. One writer devoted an entire chapter to it naming it, *'Sucuriji gigante*, a super snake of the Amazon apparently distinct from the common or green anaconda.

Other professionals in their field have suggested that perhaps an un-described and truly huge species of anaconda was responsible for the early accounts - a Pleistocene relict already on the way to extinction. It is conceivable that the early South American explorers were baring witness to an already rare and disappearing species. The ages of snakes in the wild cannot be reasonably estimated beyond their third or fourth year of life. Generally, a snake that is large for its species is old, but just how old is nigh on impossible to determine. Herpetologists do not know enough about natural death rates from predation, disease, and old age to predict how long a species could live in the wild. The only estimates available to us, of life spans of anacondas are data held on captive snakes. It is not known for sure whether any wild snakes would exceed our maximum known ages for captives. In the case of anacondas, which have virtually no natural enemies, it is possible that they might survive longer in the wild than zoo records have shown. The oldest captive individual anaconda on record is a female, which lived for more than 31 years at the Basle Zoo, Switzerland.

Our meal that night was more in keeping with camping in the open plains: corned beef and salt crackers, washed down with water. I ate my bowlful and took more of the painkillers in a bid to get some restful sleep. That night in my tent, I tended to my injuries, namely my foot as it now appeared to have a huge blister, some four inches in length covering the area I had damaged in the fall. I resisted in slicing the blister open as there was an increased risk of infection given the location. I was not overly concerned with my right thumb as it had, in the most part, become dark purple, lifeless and numb - unless I tried to use it - then a searing pain shoot up my wrist and arm. This was not too much of a problem as I was left handed. I drifted off to sleep that night listening to what I thought were rodents scurrying through the undergrowth maze of grasses near to my tent.

We all awoke just before sunrise the next day, in the hope of catching anacondas moving from the water to bask in the early morning sun. While trekking, up to our knees at some points, across the soft marshy ground, we came across a number of trails - the largest of which I would say was around 15-17ft in length. Kenard and Damon went ahead of us, and Kenard spied a young anaconda of around 4ft in length, but on being disturbed it quickly slithered into the water. Sadly, we could not locate any larger specimens curled up basking in the early morning sun. They must have heard us coming! But having said that, we could have literally been walking over them in what was the remnants of the swamp ground, and would not have known.

We could not be as fortunate as Matthew French Young, who - while on one of his many excursions in the interior via boat - saw an immense thing sliding into the water. His travelling companion at that time was a doctor, who had also seen something from the corner of his eye. He asked French Young what is was, and he replied that it was an anaconda. His doctor friend refused to believe this, saying that it was just impossible. At this, French Young instructed the boat to be moored as close as possible to the river bank. They both went ashore, French Young taking the lead. He soon arrived at an area where the bushes had been pressed down as if by some immense weight, measuring an area of around 200ft. He estimated the snake to be in the region of at least 50 to 60ft and having a girth of 5 to 6ft. Only on seeing this sight for himself did his doctor companion believe that what he had seen slipping into the water was indeed an anaconda of enormous proportions.

In French Young's book, *Guyana the Lost El Dorado*, he states, "*Records in Brazil prove that anacondas measuring some sixty feet have been known to slide into an Amerindian hut near the river and carry away a victim whose screams and disturbed water were the only signs of what had transpired.*"

We carried on with our own search. We traversed over various landscapes; its crazy this place; one moment you will be walking through long grasses, and the next over charred, baked lumpy earth, then before you even realise it, you are up to your thighs in swampy marshland. We came across a creek in our search for anacondas, and we thought we may spot a caiman or two here as it was rather secluded and an ideal spot for basking, but alas no caiman were to be found. Along the water's edge Kenard showed us a fruit, known as lanna, which was used for tattooing. He cut one of these fruits in half, and rubbed it flesh-side down on the back of our hands, and he said if you want it to be permanent you prick your skin with a needle. It took most of the day for my hand to turn black. He told us that it would fade away in 3 to 5 day's time.

All this walking on various terrain was beginning to take its toll on my foot again, but thankfully the group had decided to return to camp as the sun was getting higher and rather hotter than usual. On the way back, Damon showed us a fruit known as the savannah cherry, of which there were many varieties. This particular type was a small yellow fruit, which you split open and then just suck on the black coloured flesh, and it was rather refreshing.

Kenard by now, was ahead of us looking for anacondas as we thought, accompanied by his bow and arrows. He had told me that morning that the bow was made from wood which can only be found in one place in Guyana, which happened to be on the mountain range where the burials were!

We watched him in the distance, then suddenly he disappeared, our spirits were raised - had he found an anaconda? He then reappeared and beckoned us over, and we did our best to quickly head over to him. Kenard had, in fact, killed a young caiman. It was around 3ft long, and Richard estimated it to be around 2/3 years of age. After Damon had checked to see what sex it was (it turned out to be female), I asked if I could hold it, wanting to feel the weight, power and texture of it. It was indeed very muscular and heavy. As I held it, it gave a few final death spasms. Kenard seemed proud of his achievement, announcing, *"We shall have roast tonight"*. However, I sensed mixed views from within the group; in particular from Richard as crocodilians were, after all, his favourite creature. I, on the other hand, was looking forward to roast caiman, instead of corned beef and crackers. I also felt that it would have been rude of us not to partake in this feast, as Kenard had - after all - made a clean kill, shooting it through the head just above the eyes with one of his metal-tipped arrows. The man obviously possessed exceptional hunting skills, which he had developed over his lifetime.

We finally got back to the camp around 10.30ish, and no sooner had we got back, Kenard was suggesting that we move there and then, in a bid to avoid mosquitoes, near to another pond a few miles away, but the heat was getting the better of us, and we decided to stay and move in the cool of the afternoon. Jon and Dr Chris decided to walk with Kenard to the Ranch - a distance of around 2 miles.

Mr Biffo and I shared the makeshift shade that the guides had put up. The heat was just too harsh, and we could not take shade in our tents as that could only be described as getting into a plastic bag that was inside a microwave that was permanently switched on.

While under the makeshift tarpaulin shade, Mr Biffo and I chatted. He asked how I was feeling, and suggested I get medical attention as soon as possible. He seemed genuinely concerned for me, and I for him, as the heat and conditions were beginning to take a toll on him also.

I explained that I did not want to miss out on the expedition, it was too important. I am quite sure he thought that I must have been suffering from heatstroke, or was just plain mad, as he seemed taken aback at how well I was coping with my injuries. The heat that day was unbelievably hot; even Foster - one of the main Amerindian guides - expressed how intense the heat was. In fact even while in the shade, my back got sunburn through two layers of clothing - one of which was a UV protection shirt. Damon came over, and offered us a bowl of 'Farine', (cassava by any other name!) with sugar water. It was extremely sweet and very filling; I could not eat it all. Mr Biffo and I became rather concerned that Richard had been missing for some time, and at first we thought he may have gone off by himself in search of anacondas or caiman, but we soon dismissed this thought as we were sure he would not do something so silly.

Sometime later he emerged, and proclaimed that he had spent the last 2 hours plus up to his knees in a small pond with a small bush for shade in a bid to cool down. Shortly after 2 o'oclock, we began to break camp and move on, but the thought of this trek filled me with dread. I don't know whether the heat of the day had anything to do with it, but I was in a great deal of pain, and found it very hard to dismiss it. Jon, Dr. Chris, and Kenard re-appeared with a crate of much appreciated warm fizzy orange pop, which was duly shared out among the group, before we headed off to our next destination which was an area known as Cashew Pond.

This trek really took it out of me. When we asked how far away it was, we were told that Cashew Pond was, 'only a mile away'. I am quite sure these Amerindians have no concept of distance, and that they just walk until they get where they are going. It turned out our next campsite was more like four miles away. The uneven terrain took its toll, and I soon started to lag behind, until eventually the rest of the group were nothing more than small dots on the horizon.

When I finally got to the next campsite, Damon helped me to put my tent up, as the light was fast disappearing. Kenard was full of apologies as he was not aware of the extent of my injuries, and he was feeling extremely guilty about the distance of the trek. In a bid to make up for this he gave me his bow and arrows as a gift. I told him there was really no need, but he insisted, saying that he could make another set when he got back to his village. I thanked him, and took the items. '*My son is going to love these, if I can get them through Customs*', I thought!

As darkness fell, a fire was lit so that the caiman could be roasted. We all sat watching it cook and listening to it whistle, pop, fizz and crackle in the heat of the fire. It was rather hypnotic, watching the flames dance round its stiff, lifeless body. When the time came to partake of this savannah feast, Richard pulled his face, at the thought of eating the caiman, but after much persuasion, he exclaimed that it tasted like cod. I thought this to be odd as cod is a saltwater creature, and caiman are freshwater. I was offered a leg; I happily bit into it - the taste was very much like chicken, not fishy at all. I was then offered meat from the tail which is thought to be the best bit. Now this did have the *texture* of cod, in that the pieces of cooked meat came apart in large flakes, but as for the taste, again it was very much like lean roast chicken. We did not stay out of our tents too long that night, as the insects were beginning to make a meal out of us, and to be honest I think that all of the group were just plain exhausted from the ferocity of the heat during the day.

As daybreak came round again, we broke camp and made for a place known as Point Ranch. The bus had been arranged to pick us up around eight that morning. Again this journey for me was long and hard - I could not keep pace with the group - and sure enough I began to flag behind, watching the rest of the group disappear as little dots into the haze that was the early morning horizon.

Upon arrival at the ranch, we bought pop and beer from the old couple that lived there. While chatting, they offered us fresh-made coffee. I must say this was the best tasting coffee I have ever tried. While partaking of this hospitality, the old chap, Elmo - who owned the ranch - began telling us stories of water tigers. He had seen them many years ago.

He explained, and was quite adamant, that they were not giant otters, as he was familiar with these animals. He went on to say that the water tigers he had seen were spotted very much like a jaguar, but instead of being a solitary hunter, they hunted in groups. He also stated that there was a Master (this may be a parent) that sent the younger ones ahead. He had seen such a group a number of years ago on a local mountain, and he turned and pointed to one behind him in the distance. This mountain had no known name, but it was well known that a dragon (quite possibly a large snake, an anaconda who knows?) guarded a spring which was located on this mountain. Many people had climbed this mountain, but not a single person had ever returned.

Kenard added to this conversation, saying that it was well known that water tigers were of various colours and patterns, namely spotted, brown, and white with dark spots. Joseph, another guide in our group, said he had seen an actual hide of a water tiger which had been killed by a hunter in the 1970s. In total it was some 10ft in length. The patternation was white in the most part, with black spots (at this juncture he turned and pointed to a number of cows over in a field). The head was still attached to the rest of the hide, and he stated that it was striped like a tiger skin.

As we listened to what the locals had to say, it struck me that this creature may actually be a form of unknown giant mustelid, as certain species such as stoats can, in fact, change the colour of their coats, and they are known to hunt in family groups (mink are also known to hunt in groups). We took another break drinking our warm orange pop, before Joseph then approached us. He had another story to tell.

In the mid-70s, a plane crashed in the local mountain range. Being a ranger at the time he was asked to trek out and retrieve the body of the pliot. After a day's trekking in the mountain jungle, he found the wreckage and the body. The head was missing, and the body had been badly burned in the crash fire. He recovered the body from the wreckage, and wrapped it up, and placed it about his person in a sling-type manner. Then began the return journey, through the mountainous jungle. However, he became lost, and for 3 days he wandered in the mountainside jungle. While lost, he had no choice but to eat the charred flesh of the crash victim. He nibbled at the arm of his dead host, in order for himself to stay alive. He finally found his way out, and returned the body minus one head and a bit of arm. I could not help wondering how he explained the teeth marks!

At that point the bus arrived to take us back to Lethem. As soon as we got back to town, we made arrangements to fly back to Georgetown, as none of us could face another 18 hours on a dirty, hot sweaty bus.

While we were there, the guys made some enquiries about hiring a helicopter in a bid to get to Corona Falls. After all, this is why we had come here to possibly see outsized anacondas. But there was only one, and that was not available. The other option was to hire one from over the border in Brazil, but the downside to this, however, was possible days of expensive red tape with no guarantee of being granted permission. A boat was out of the question, as Kenard told us the river was too low. I had had my reservations about coming in the dry season, and all this frustration about getting to

Corona Falls seemed to prove my point. Richard decided to try and return next year during the rainy season, and charter a boat or plane in a bid to get to Corona Falls.

While sorting out plane tickets, we were told by Damon and Kenard that while the new landing strip for Letham airport was being constructed a number of termite mounds were cleared. As they were removed, it was discovered that at least 11 of them were found to contain what appeared to be human remains in a crouched position. Being in this position suggests that the mounds may have been broken open, and the bodies of the dead placed in there, or another option for consideration it that the bodies may have been placed in crudely built stone cists, and over time the termites may have built their nest around them.

These burials were common knowledge among the locals as, when asking around, they all seemed to recall the discovery. However, neither Damon nor Kenard knew of any Amerindian tribes that had a tradition in which the bodies were dealt with in this manner. Damon was of the opinion that these burials could well have been pre-historic. I am sad to say that the bones from these burials were thrown away and no record was ever made of the them. I have never heard of such a burial style and, who knows, there still may be one or two of them left. To find such a thing intact and undisturbed would be somewhat of an archaelogical coup!

Again we checked into the Tauku Hotel. I could not wait to freshen up and tend to my wounds. I sat on my bed looking at my blister engorged foot. I could not take the pain any more, or the fact that I felt as though I were slowing the group up while out in the field.

I took my knife and gently cut into the huge fluid filled sack of dead opaque skin. It was a wave of relief as the yellowy/red liquid oozed out, and I could physically feel the intense pain drain away - it was bliss. Just then Richard knocked on my door to see how I was getting on, and as I turned to look up to say *'much better now'*, my jaw dropped in horror. Richard was wearing a T-shirt and shorts and of the skin I could see, he was absolutely covered in bright red lumpy bites, so much so he looked as if he had small-pox. It turned out that he had been bitten rather badly by vine flies.

I took pictures of his bites just in case his condition got serious, and in return he took pictures of my injuries for er…. posterity. Cutting away the large amount of dead skin on my foot had revealed a number of deep and extremely tender lacerations on the sole which must have happened when I had fallen off the mountain.

That evening we sat on the front veranda of the hotel and watched a herd of horses stampede through the hotel grounds and then disperse on to the main road. Nobody seemed too concerned. Later Kenard told us stories of the di-di - of how a woman had lived with them under duress, bore a child by one of them, escaped via swimming a river, and watched in horror from the opposite river bank as the di-di tore apart this hybrid child.

The next day after lunch, Damon had arranged for us to meet with a former tribal chief who knew a great deal about the many strange creatures of Guyana. We drove along a long, dusty track some distance outside Letham town, and again slowly we saw the landscape change from sparse arid dryness to lush deep jungle greens of foliage at the base of the Kanaku mountain range. We then drove up a twisting jungle path to a clearing near a stream. There, set in the beauty of the light jungle, was a middle-aged man waiting to great us. He was wearing of all things a 'Sideshow Bob' T-shirt, which I found most amusing. His name was Ernest. He was now a retired tribal chief, preferring to spend his time setting up his own small fish farm. He took great pride in showing us his meagre fish stock.

Ernest was a wealth of information regarding the creatures we had come to learn more of. He started by telling us that some 10 years ago he had seen an anaconda of around 30ft in a pool some 20-plus miles away. Sadly, he then went on to say that an Englishman had shot it dead, and it was thought to have been transported to England. Of course, this would have been done illegally, but wealthy people have their ways and means to preserve their barbaric notions of fun.

He had heard of the water tiger, and recalled, while out fishing on the river with his Grandfather, some 20-plus years ago, something seizing the small boat from beneath, and whatever it was began shaking it violently. His Grandfather told him that it was water tigers, and to quickly hold on to the branches of the trees that overhung the river in a bid to stop it overturning while the boat was rocked violently from side to side.

Ernest was also familiar with the little red faced pygmies that we had come to hear so much of on our travels. When he was around 19 (he is now 59) he had seen one. It was naked, brown skinned, and had a red face. Unlike Kenard, he believed that the red face was natural and not paint. The little man, without a sound, had taken tobacco from Ernest, before then disappearing back into the forest. Again he re-affirmed that they like tobacco, and he was of the view that they were not dangerous unless in danger. One of the ways in which they become angered is if you encroach on their space. Their homes are made under large trees. *"If these were cut down"*, he said, *"they become, naturally, quite angered"*. Earnest said that on occasion, he came across very small pots on the jungle floor, that the red-faced pygmies had made, and he warned that these should be left alone. You could not engage in conversation with these pygmies - they just simply would not respond to you. They seemed to take tobacco and leave, disappearing into the deep jungle.

At this juncture, Damon confessed that he had seen a pygmy as well. A number of years ago had been camping with his sister-in-law and another girl. He awoke in his tent to see a tiny red-faced man grinning down at him. He said he was was frozen with fear. In time he found that he could move just enough to nudge the girls awake. But it was too late, for when he looked again the pygmy had gone.

Earnest said that he had not seen a di-di himself, but one of his friends had. However he'd died some two years previously. Moreover, it had been a female he had seen, which - at the time - was in a tree, suckling an infant. He observed them for some time before, he thinks, blacking out. On regaining consciousness he became continually ill. His illness became steadily worse, and this he blamed on the di-di sighting. He only revealed this information while on his deathbed to his daughter. Ernest said that the di-di sounded very much like a human shouting, and from time to time the long deep shouts could still be heard in the Kanaku Mountains.

His final story really captivated us. In fact none of us, including Richard, had heard anything like it. A couple of years ago in a little cave at a place called Wa-sa-roo Ernest had seen what he took to be a tiny caiman. It was brown in colour and had a red strip running down the length of its back. This enthralled us, even more so Richard, as this description does not match any known species of caiman. The impressive bit was yet to come - Ernest said that these tiny caiman had two tails!

Now, I am aware that you can get the odd genetic hiccup, as is the case with snakes or lizards etc, with two heads. Or, could it have been that Ernest, had in fact seen two of these mysterious caiman mating, or maybe just one simply laid on top of another while estivating. There's even the possibility that the 'red stripe' gives the illusion of two tails as a form of defense. Ernest did add that while he was watching these small two-tailed caiman, he could hear a loud, deep bellowing type of noise. He said the sound did not match what he was seeing.

It was time to move on, and we thanked Ernest for his time and his information, and we then made our way to Wa-sa-roo, to see if we could see for ourselves this unknown caiman.

The only way I could possibly describe Wa-sa-roo, is to say it was straight off the set of *Indiana Jones*. There was a huge collection of boulders - some of them as large, if not larger, than an Amerindian hut - as we neared the entrance of the cave, and set well in amongst the boulders we could hear running water. The boulders when they had fallen, had fallen on to the path of a stream, which had been transformed into a small waterfall inside the cave. Richard was straight in there, closely followed by me and camera. With the blister sliced away, I could move with ease, and was back to my old self - scrabbling all over this mass of boulders, vines, and roots. There was only room for one person to get through the entrance at a time. I gave Richard my camcorder, and switched it to night-vision, so that it could pick up any potential eye shine of creatures within the dark of the cave. He entered, and shouted back that it was cool inside - just right for caiman to estivate, although he could see no direct evidence, such as tracks or nesting material, but nonetheless it was an ideal place for caiman. It had all they required - water, food, shelter and remoteness.

We then turned our attention to climbing over the top of the boulders, as another route in via the top had been found. I used the night vision on my camera again, in order to pick out anything of interest. But not a great deal could be seen, apart from my camera picking up a rather unusual cave cricket.

Damon reminded us that time was getting on, and we had to make a move back to town, as we had been invited to Ernest's family home for an evening meal. While walking back to the truck, Damon called me, and he showed me what appeared to be an oversized fly. It landed on his leg and then proceeded to draw blood from him. It was in fact a caboura fly (possibly Similium spp. Also known as blackflies). I quickly took a number of pictures as it was amazing, watching it draw and then dine on Damon's blood. After a few moments of this, Damon did ask if I had all the pictures I needed, as he felt it had drank enough of his blood for one day. On the return journey to the hotel, the driver and I spied a Neo-tropical rattlesnake slithering off the dirt track, and disappearing into the scrub savannah. It turned out none of the guys saw it as they were all facing the other way in the open truck watching the mountains grow smaller and smaller.

That evening we dined at Ernest's family home, which was just outside the main town of Lethem. The conversation was interesting to say the least, and Damon told us that a number of his family had, in the past, worked at a Canadian-owned open cast gold mine, and at one time it had been the largest in Guyana. A large number of the workers had been witness to the discovery of a very large human-like skull. Damon had pondered over it many times wondering if it belonged to Gigantopithecus. As far as I am aware, the only remains of such a creature have only ever been found in China and India. The story ran that officials took the skull away; it was never heard of, or seen, again. For all we know it could be mounted on a plinth in some millionaire's study, or even locked away in an office basement to be deliberately forgotten about.

Foster made mention of an even more bizarre story. A number of years ago, a man-like creature with webbed digits had been swept into a village during flooding in the rainy season, the best way he could describe it was to say it looked like the *'creature from the black lagoon'*. What this creature was remains a mystery, as details were rather thin on the ground. However, Damon did say that the story had made it into the local paper. Mr. Biffo said that there was an online archive of the local paper, and he offered to check it out once we were back in England. At this juncture of the evening we were joined at the table by a very large cane toad, which had hopped in. Richard picked it up and I took a couple of snaps of it. We had a very enjoyable evening and thanked our hosts for a most hospitable night.

The following day, we caught the tiny plane back to Georgetown. Damon and Foster had to stay in Lethem as they had to return to a remote village to collect some snakes and invertebrates that had been gathered for them. While at the tiny airport a truck pulled up, which - even when the engine had been turned off - was still very, very noisy. It turned out that the truck was full of green winged macaws. They were scheduled on our flight to end up God knows where - it was awful to see them; I went over to check them out. I counted 18 of these magnificent parrots cramped into a box that was, at most, 3ft long. I was so angry at the sight of this, for every bird that reaches Europe alive, another 3 or 4 die in transit. You do the maths!

Time passed, and it was time to board our flight. Now, I am used to the simplicity of small planes, as my Uncle used to build and fly his own. Mr. Biffo, however, was not, and to say the least he was extremely nervous. He had never experienced such a small aircraft; he likened it to a 'tin shed with wings'! I was quite happy, and took my seat right behind the co-pilot; I got my camcorder out in order to film the take-off, and the landscape below. I did my best to make Mr. Biffo laugh, but the plane did that by itself. As we took off, the back of the pilot's chair fell off into Mr. Biffo's lap, as he had chosen to sit behind the pilot. In a way it kind of justified his nervousness, and did little to calm his fears. His facial expression of sheer terror was priceless; we just all looked at each other and began to cry with fits of laughter. What can I say? It was just one of those moments.

On reaching Georgetown, Damon's nephew Maradonna was waiting to greet us. He was to be our guide for the last leg of our journey. Mr. Biffo, by this time, had had enough of camping, insect bites, blisters, and self-destructing planes. To top it, off he was suffering the effects of heatstroke, not to mention the fact that he was missing his family terribly. He wanted to see if he could get his flight home brought forward. We took him to the main airport, and he found out that he could at least get to Barbados, with the possibility of moving his flight to Gatwick forward. He was not sure whether this could be done, but he chose to risk it. The flight was not until 3 am, so he arranged to stay with the taxi driver, a chap called Marvin.

For Richard, Jon, Dr. Chris and myself, the next few days were spent as guests at the Pakuri settlement. Maradonna proudly showed us round the settlement. The local children, with pride, showed us their collection of animals they had gathered for Damon - a small peccary, and a couple of very young caiman that the boys had come across while out swimming.

We stopped at a few bars en-route. I asked why there were so many, and why didn't they just go into town? He explained that there is great racial tension between the Blacks, Asians, and Amerindians, and that it was very rare that they would travel into town, as more often than not, it would end up in a fight, and the Amerindians would always be the ones to be arrested, as they are looked down upon, and thought of as wild and uneducated by the Guyanese. In actuality, nothing could be further from the truth.

When we returned from our 'mini tour', Richard sat chatting to Foster's father, Joseph. He said that prior to the settlement being built, the area was well known for the 'little red-faced men'. However, he also said that they would like to wrestle with people, and that the way to defeat them was to knock them over to the ground, as he said their legs did not bend! Make of that what you will. Joseph also relayed that he has seen a large anaconda around 5 years ago. He estimated it to be 23ft in length. His father had told him many years ago that the water tiger had once lived in the vicinity of the settlement. It dwelled in a cave on the riverbank. He described it as being brown, and rather dangerous.

Later that evening we were shown to our room, in which all four of us were to sleep. It was a traditional Amerindian house on stilts, with a thick roof of what looked like palm leaves, and as I unpacked I spied a beautiful full grown pink toed tarantula, nestling in the corner of the roof. I showed the guys our extra roommate; they duly took snap shots of it.

That night we watched Guyanese television. This was a relatively new concept as prior to the generator that was installed a couple of months ago when it went dark, that was it, so to speak. Juanita, Foster's daughter, told me that since the generator had been installed, the number of incidents with Neo-tropical rattlesnakes had also gone down. Maybe the vibration was having a marked effect on the local fauna. Guyanese television leaves a great deal to be desired, I must say. There was a programme on racial unity in Guyana, and at no time was there any mention of the Amerindians. Every now and again you heard a little snigger or tut from Joseph, regarding the remarks being made on screen.

The next morning after bathing in the reddy brown waters of the creek (which, incidentally, had risen quite a considerable amount from the previous visit, so much so that there was a boat of fishermen held up there sorting there nets when we arrived) Juanita had prepared a fantastic breakfast of fresh made rotia, which is a type of bread, fried green beans with what appeared to be hotdog sausage, fried eggs, and coffee. The bread tasted amazing.

Later that day the guys went for a canoe ride along the creek to see if they could spot any interesting wildlife. I stayed at the house, as my feet - or rather my toes - we hurting as the nails on a number of my toes were in the process of falling off. Juanita offered to bore holes in the nail to relieve the pressure, but I politely declined. Instead, we chatted about the state of affairs regarding the Amerindians, and it turned out that she is a school teacher, and was hoping to get a job at the newly built school that had been erected in the village. She told me that there was great political debate over where this school should be built, as even their own Minister for Amerindians didn't want them to have it. We also talked about more sensitive issues, of how very young girls are lured by unscrupulous Guyanese into jobs that don't exist, only to find themselves sold into the sex industry. The response from the authorities is appalling; they genuinely do not care, if a victim turns out to be one of the 'Bucks'. This is how the indigenous people are referred to by most Guyanese. Juanita explained this racial slur - it implies that the Amerindians are wild and beyond taming, like the deer. My personal experience of the native people is the complete opposite. They are highly intelligent, and very socially organised within their own communities, and have a phenomenal understanding of Guyanese flora and fauna.

All too soon our short stay with Foster's family came to an end, and we travelled back to Georgetown and checked into a spotlessly clean guesthouse. It was quite a shock really after spending all that time in expedition mode. It was situated in a more pleasant part of Georgetown, in fact just round the corner from the Prime Minister's residence. Maradonna told us that the zoo was not too far, and he would escort us there and back. At this juncture may I say that I have always thought you could judge a society by the way it keeps its animals in the local zoo? Guyana Zoo for me was THE lowest point of the whole expedition - it was totally appalling.

The animals were kept in horrendous conditions. As we walked round we bore witness to green anacondas in a dry concrete cage completely void of any water. At one point Richard spied a stop tap round the back of the pen, and jumped over the tatty fence, in a bid to give the snakes some water, but the tap was broken and we were unable to help them out. To my horror, in between two cages a makeshift pen had been built, and housed in the dirty polluted water, were at least five, maybe six, dwarf caiman, in a space that could not have been more than 1½ft wide, by around 6ft long. I just could not believe the blatant disregard for the well being of these creatures. One

of the poor things had had the top of its jaw bitten off, leaving the lower internal jaw exposed.

I suggested to Richard that we needed to film this terrible sight, and make the relevant authorities aware once back in England. This filming had to be done covertly, as there were signs saying no filming unless you had paid to do so. We had no intention of paying, as it was quite evident that the money would not have gone to the upkeep of the animals. By far one of the worst sights was a full-grown African lioness, which was housed in a concrete cage no bigger than 12ft by 7ft. This once proud creature had been reduced to rocking to and fro, displaying the classic signs of stereotypical repetitive behaviour. On closer inspection, it also had sores on its rear pads due, no doubt, to only ever walking on concrete. Across from this cage, a group of Guyanese 'keepers' (I use that term very loosely) lay lounging on the seats in the midday sun, not giving the animal a second glance! I glanced at my watch - it was around 1.30 in the heat of the afternoon, and I witnessed a beautiful jaguar being let out of its night pen - shocking! It made a dash for its concrete bowl, and quickly drank what few drops of water it had. We came across a man-made lake; it looked more like an open sewer, with pop cans, paper bags all entwined in a thick oily film. This was the manatee enclosure! I just could not believe my eyes when I saw, on the far bank, a number of men fishing in the manatee pond! That was it for me. I could not bare to stay in the godforsaken place any longer.

Richard has promised (or rather sees it as a personal crusade) that he will do all he can to make the relevant authorities aware of the terrible ignorance and appalling treatment of animals in Guyana Zoo, and hopes that this sorry state of affairs will be addressed.

After the zoo experience, nobody was really in the mood to do much else, so we retired, and had an early night. The next day we visited two museums.

The first one we visited was the Walter Roth Museum of Anthropology. Now, I was looking forward to this, to see if any information could be gleaned regarding the 'termite' burials. However, the place was kept in a similar condition to the zoo - poorly laid out - and it looked as if its upkeep and maintenance had stopped with the departure of British rule! It was badly laid out, and dates were missing from a number of artefacts, but over in the corner I saw a white man sitting in a room (the sign on the door said 'Head of Archaeology) surrounded by paper work. I approached him, stating that I was a keen amateur archaeologist myself. I asked had he ever come across human burials within termite mounds. At first he seemed very interested and keen to learn more (he also sounded French though I could not be sure). I went on to say that at the time it had been common knowledge among the locals (I did not divulge the location) as we had asked a number of them independently. He was beginning to open up and be rather chatty, as I had captured his attention with the mysterious burials. He then asked who our guide had been, and we told him. His attitude changed immediately, and he gave a loud *"HA!"* *"Damon Corrie"*, he said in raised voice, *"You don't want to listen to what that fool tells you, his information is wrong. Anyway what would he know, he's not from Guyana, he's from Barbados!"* I could not believe my ears. I replied with: *"Well, Sir, you don't look or sound as if you are from Guyana yourself, and may I say you keep your museum in an awful state".* On that note I promptly walked down the stairs to place some distance between this rude man and myself, so that I did not punch him and land myself in a Guyanese police station. As I stepped out of the door I shouted, *"The ignorance of these people!"* I just could not believe what I had heard. Richard followed me out shortly afterwards, saying that he too had also had a dig at the curator.

The next museum we visited was just as bad. I can only equate it to a very shoddy second-hand tat shop! It had the most bizarre collection of stuff I have ever seen. On the ground floor was a collection that can only be described as souvenir dolls, mixed in with an array of toy cars. In the case next to these was a group of random objects including a knuckleduster!

Upstairs was even worse. The collection of poorly stuffed animals seemed to be held together by dust and time itself! Most of them were the wrong colour due to sun bleaching. Over in the far corner hung on the wall was a stuffed anaconda, and Richard managed to get some film of it, before being told abruptly to stop filming by a member of staff.

For me the two museums and the zoo seemed to be a mere reflection of how badly run Guyana - as a country - is. During our short stay downtown, I yearned to be back out in the field as it were, looking for outsized anacondas and mysterious red-face pygmys, instead of wandering in this man-made hellhole, searching for signs of humanity.

We were more than happy to leave Georgetown for the more sedate island that is Barbados. We had 10 hours to spend there before our final connecting flight back to England.

We were met by Damon's father. A truly gracious, wonderfully mannered gentleman, and, of all people, Mr. Biffo!

He had not been able to bring his flight forward, and the jammy bugger had been stuck in Barbados for the last three days. He told us a dark scary tale of how

 **CFZ GUYANA EXPEDITION 2007**

Marvin's house was nothing more than a shanty-like hut void of any glass, and situated in a delightful position, next door to an open sewer. On entering this abode, he could hear and see small shadowy creatures scurrying about in the darkness of the corners. As quick as a flash he offered to accompany Marvin on his taxi rounds in a bid to avoid being some creature's potential dinner! He told us that while out in the cab, he got a sense of sheer violent tension just waiting to explode on the streets of Georgetown. At one point, he said the van became surrounded by men all screaming and banging on the windows. He was not too sure why.

We were as pleased to see Mr. Biffo, as he was to see us.

We were taken back to the Corrie home to meet Damon's mother, brother, wife, and children. A strong resemblance ran through all male members of the Corrie family. On his mother's side Damon is descended from an Arawak princess. Her son had handed over the chiefdom to Damon upon his death in 1998. Damon could also trace his relatives back to Scotland in the 10th Century! All this information was displayed with pride in neatly framed photos and paintings.

A mini-tour of the island had been arranged for us. Barbabos truly was a paradise, even more so after spending time in the depressing filth that was Georgetown. Everything about Barbados was exactly how it's portrayed in the media - beautiful vivid blue seas, soft white sand, and lush palm trees gently swaying in the breeze. Oh! How I now wished that I had taken the chance of bringing my flight forward with Biffo and failing, only to be held up in this dreamy heaven.

One of the stops on the whirlwind tour was to see the Chase Tomb, famous for its 'creeping coffins'. This tomb, or small crypt, is situated in Christ Church Graveyard. It is constructed from large cemented blocks of coral, and its dimensions are approximately 12ft by 6ft; quite a compact space. On the approach you descend a few steps into the chamber.

Nothing seemingly happened out of the unusual for the first two burials - Mrs Thomasina Goddard (31st July 1807) and an infant, Mary Anna Maria Chase, buried 22 February 1808. However, on the 6th July 1812 the tomb was opened to bury Dorcas Chase. With the door being heavy, it took several men to open it. What they found filled them with unease and fear. The two coffins previously entombed, were standing on end against the wall. Now both these coffins were encased with lead so some force would have to have been applied in order to do such a thing, but there was no sign of any disturbance. The dust was as it should be - undisturbed - and there

When the burial of Thomasina Clarke took place on 17 July 1819, the Governor of

Barbados himself, Viscount Combermere, oversaw the sealing of the tomb. Nine months later he returned to check the state of the tomb and again found it in disarray. Yet the seals on the door he had personally put there remained intact, and there were no other signs of entry. In 1820, the tomb was emptied without the mystery of the "creeping coffins" being solved. The coffins were all reburied at another location, and to this day there has not been an adequate explanation for the mysterious movements of the creeping coffins. If the movement was caused by floodwater why was there no evidence of this? Indeed, why were neighbouring tombs not affected? If an animal was to blame, why were the remains not scattered around the tomb? If indeed it was human force, why were there no footprints or signs of the coffins being moved around in such a confined space?

The mini-tour moved on, and we were shown the historic sugar cane plantations that have now been assigned to the history books, with modern day tourism taking its place as a form of income for the island. We took a brief look at a former grand mansion now in ruins, which was used for the setting of the film, *Island in the Sun*. Time was now getting the better of us and we returned to the Corrie household for a sumptuous meal of local fayre, which consisted of breaded flying fish, roast cassava, lightly battered sweet potatoes and salad. The Corrie family were truly hospitable, and I thought to myself: *"I wish that ignorant prat at the museum could see this"*. All good things must come to an end, and we had to leave to catch our plane back to the cold, damp, dark grey that is England in winter. I think, at that moment in time, we all wished we had gotten stuck in Paradise just like Mr. Biffo.

On reflection, it was a shame we could not get to Corona Falls, or indeed capture any actual footage of anacondas in the wild. But, as I have come to understand with a CFZ expedition, you come back with more questions than answers. I have come away from Guyana wanting to know more regarding these previously unknown red-faced pygmies. To me they were just fascinating, and unheard of to western ears, and no matter whom we questioned or where, the stories were consistent, with little or no deviation. Could the South American interior of Guyana be playing host to its very own living *Homo florisensis*? Sadly, as humans expand and move in, the `little men` recede and move out, moving further into uncharted mountainous jungle ranges. And of the di-di - well - conversely this may well be a larger hominid, a relation to the Canadian sasquatch or the American bigfoot perhaps? Could they even be remnants of prehistoric man? It would be so interesting to thoroughly investigate these remote regions of the interior, to see if any evidence of their habitation can be located. But, as with the pygmies, humans slowly encroach upon them, causing them to retreat into the unknown wilderness, which is seemingly more than familiar to these little-known, or understood, mysterious creatures.

With the water tiger, at first I think that we were all of the view that it may be a giant otter, but listening to accounts and descriptions it is very different. From witness accounts, it is known to socially interact within what appears to be a family cluster, on occasion hunting in packs or groups, and it can be extremely aggressive even if unprovoked. Its various colours may be an indication that it may actually change colour as it matures as other smaller mustelids do. For me, a pack-hunting, semi-aquatic flesh-eating mammal points to this creature being an unknown giant form of mustelid, possibly similar to a stoat or a mink. The last word on the expedition must go to the account of the two-tailed red-backed caiman; this was for me an absorbing description of a potentially as yet completely unknown species of crocodilian. It certainly merits its own expedition into the interior, just to solely concentrate on amassing more information of this potentially, entirely new species.

As I mentioned, we came away with more questions on information we did not even have before we had investigated what Guyana has to offer. Going on the information that we have amassed from just one expedition, it has highlighted (and I am sure you will agree) just how little is known of the interior of this vast expanse of land.

Guyana for me remains one of the world's most exciting, and least known, destinations for adventure and travel in the true style of exploration. I have come away from Guyana with a deeper and profound respect for the culture, knowledge, and exceptional skills of the true indigenous people; the Amerindians. It is a changing landscape of dense tropical rainforest, dry arid mountain ranges, and wide expanses of savannah, which is all too rapidly disappearing. The interior teems with wildlife - known and unknown - all connected and criss-crossed by a network of rivers, streams and creeks. Don't expect to find luxury resorts, creature comforts or, indeed, modern basic commodities such as roads. Be prepared to travel extensively on foot, boat, or small flimsy plane to experience the true hidden nature that is Guyana.

Not quite the Caribbean, not quite South America; Guyana, the 'Land of Many Waters', is not quite like any other place on Earth.

Alliteration time: Chris Clark in Canoe in Creek

Pakuri tree - the only one left near the village which shares its name

Crossing the Essequibo - a rickety old coach on a rickety old ferry

Top: The Essequibo
Bottom: On the track of Unknown Animals

The savannah of mid-Guyana

Top: The expedition treks towards Taushida
Bottom: Paul looking disturbingly like George Michael

The road to Taushida

Guyana: The savage land

Top: The creek where the expedition bathed, and were nibbled by little fish
Bottom: The terrain through which they walked was harsh and unforgiving

Top: The country around Taushida
Bottom: Taushida village

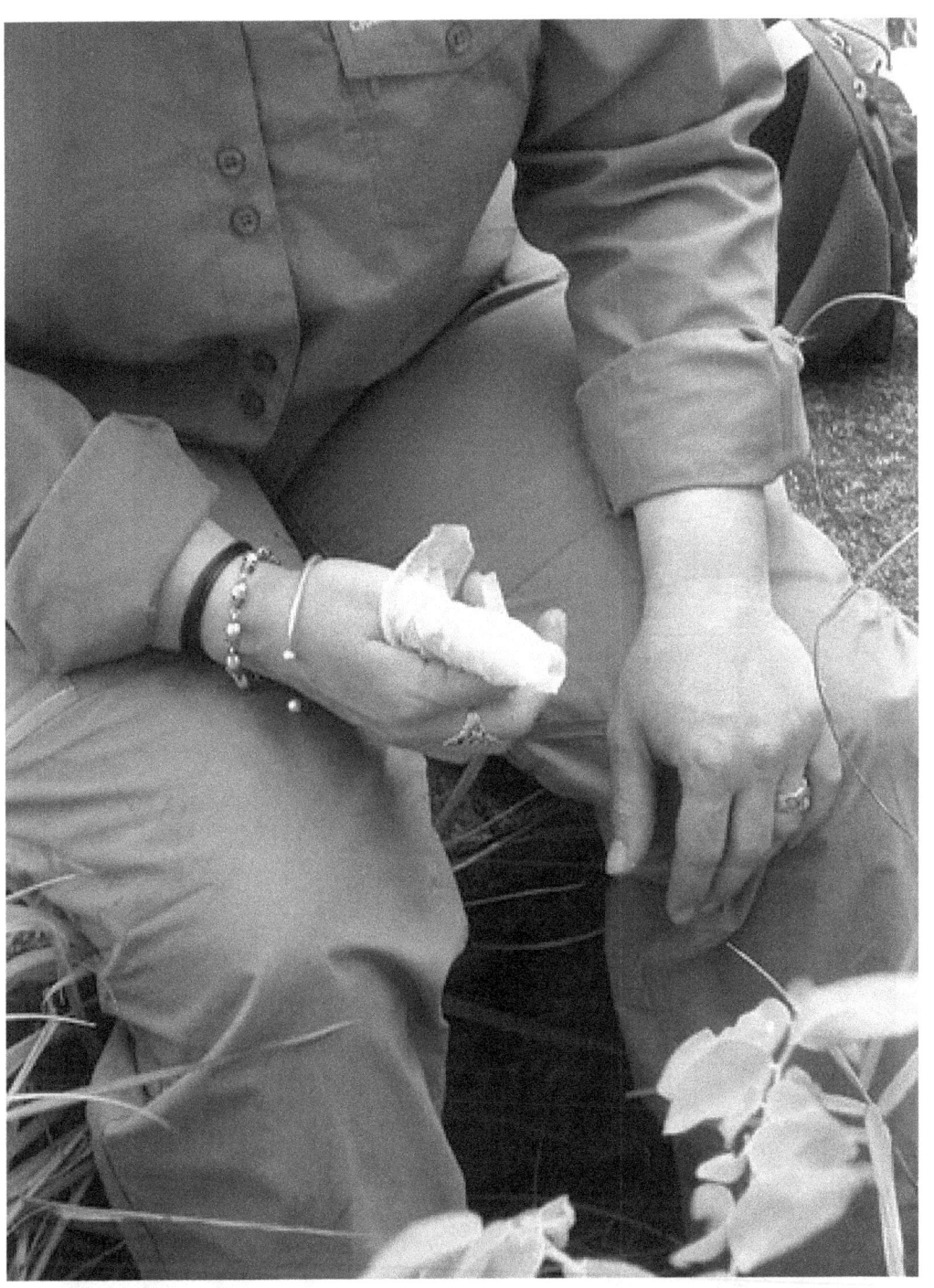
Lisa nurses her injuries with fortitude

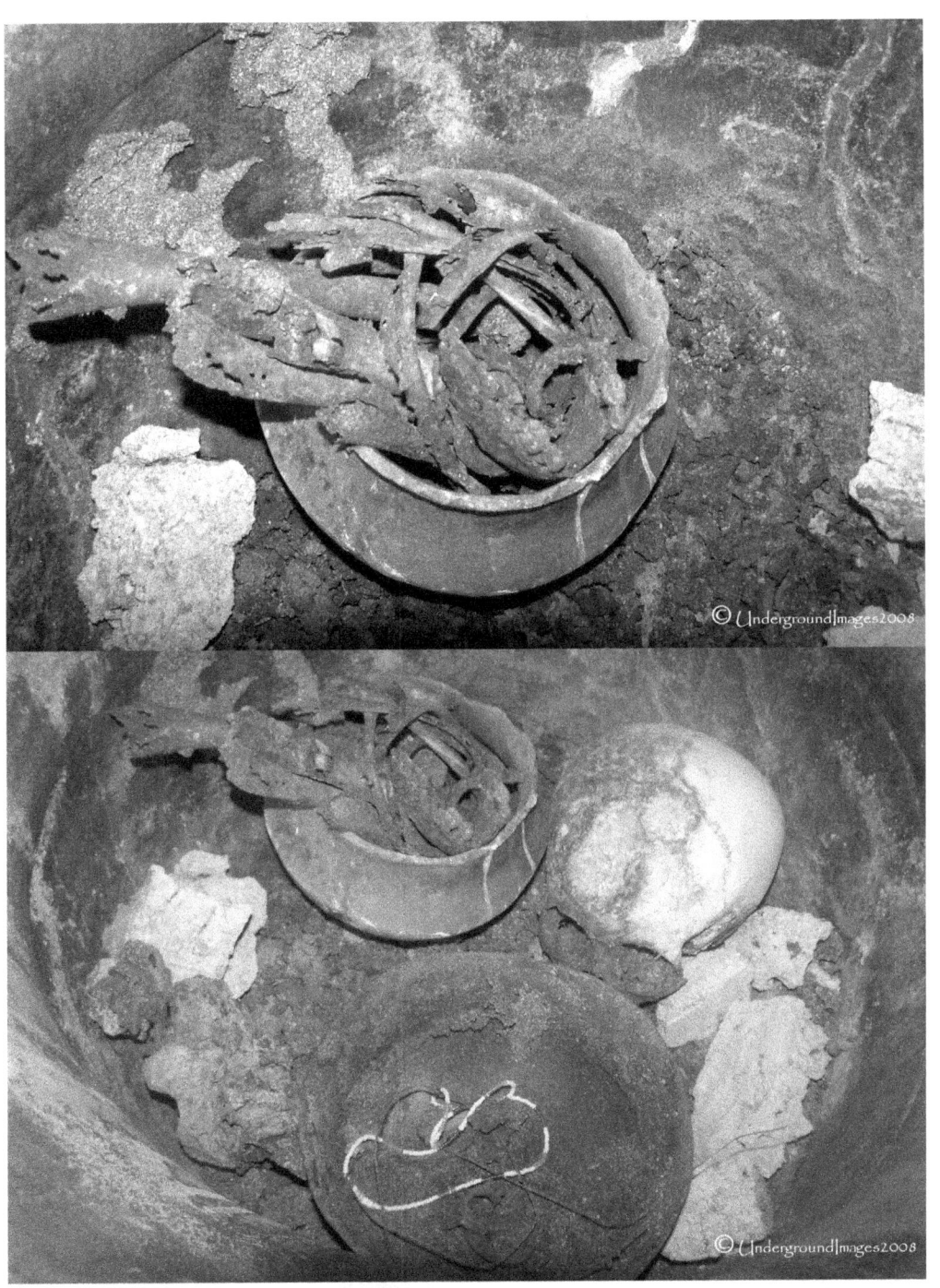

The cave burial. Note the biro for scale, and the string of beads which were left as an offering

Both Pages: The cave burial. The skull belonged to a child

Both Pages: The cave burial. The jaw is from an adult, perhaps a shaman.
Below Left: Moses Isaac who discovered the cave burial in about 2002

Both Pages: The cave burial.
Above: View from the cave mouth **Left Lower:** Damon

Both Pages: Views of Tebang's Rock

Tebang's Rock

The savannah at dusk near Tebang's Rock

Top: Crossing a creek
Bottom: Damon's brother-in-law, Foster, with ancient stone axe

Dinner

Anaconda track (insert: green anaconda).
Top left: Same view with Biffo for scale.
Below: Paul and Richard looking intrepid

Left Top: Kanaku mountains **Left Below:** Ernest (Ernesto Faris)
Above: Wa-Sa-Roo - the cave of the caiman

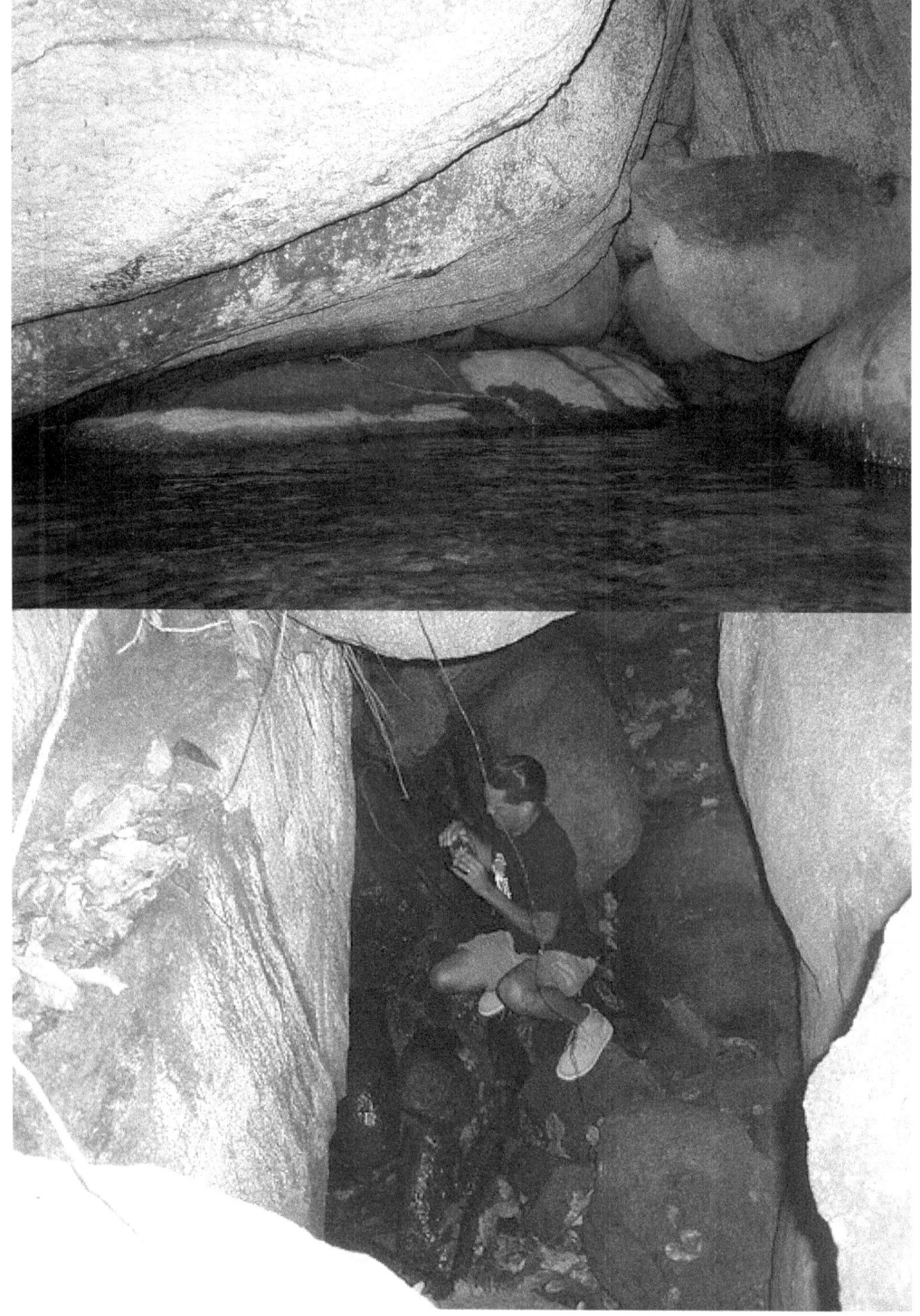

Left: Inside the caiman cave (Damon, below). **Right:** Outside the caiman cave

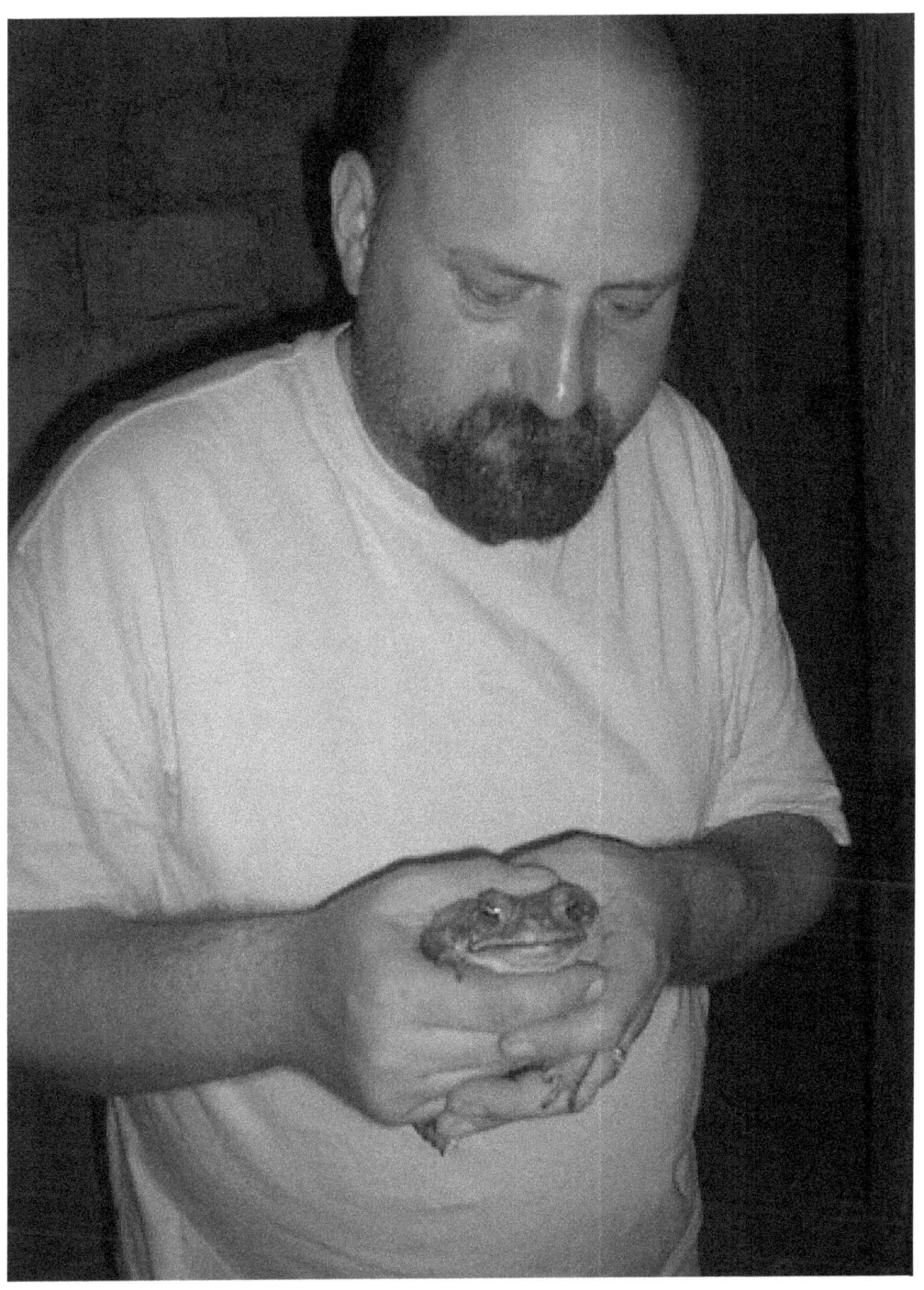

Left, top: Chris and Paul at a giant anthill **Left below:** Damon
Above: Richard with cane toad *(Bufo marinus)*

Termite mound

Top: Not a termite mound, but a peculiarly shaped rock
Below: Kenard Davis

Left top: Damon **Left below:** Cashew tree (insert: fruit)
Above: Ernest

Above: Dr. Chris
Right top: The strange colour phase of the rainbow boa *(Epicrates cenchria)*
Right below: Savannah

Above: Giant anteater *(Myrmecophaga tridactyla)*
Below: Cuvier's dwarf caiman *(Paleosuchus palpebrosus)*

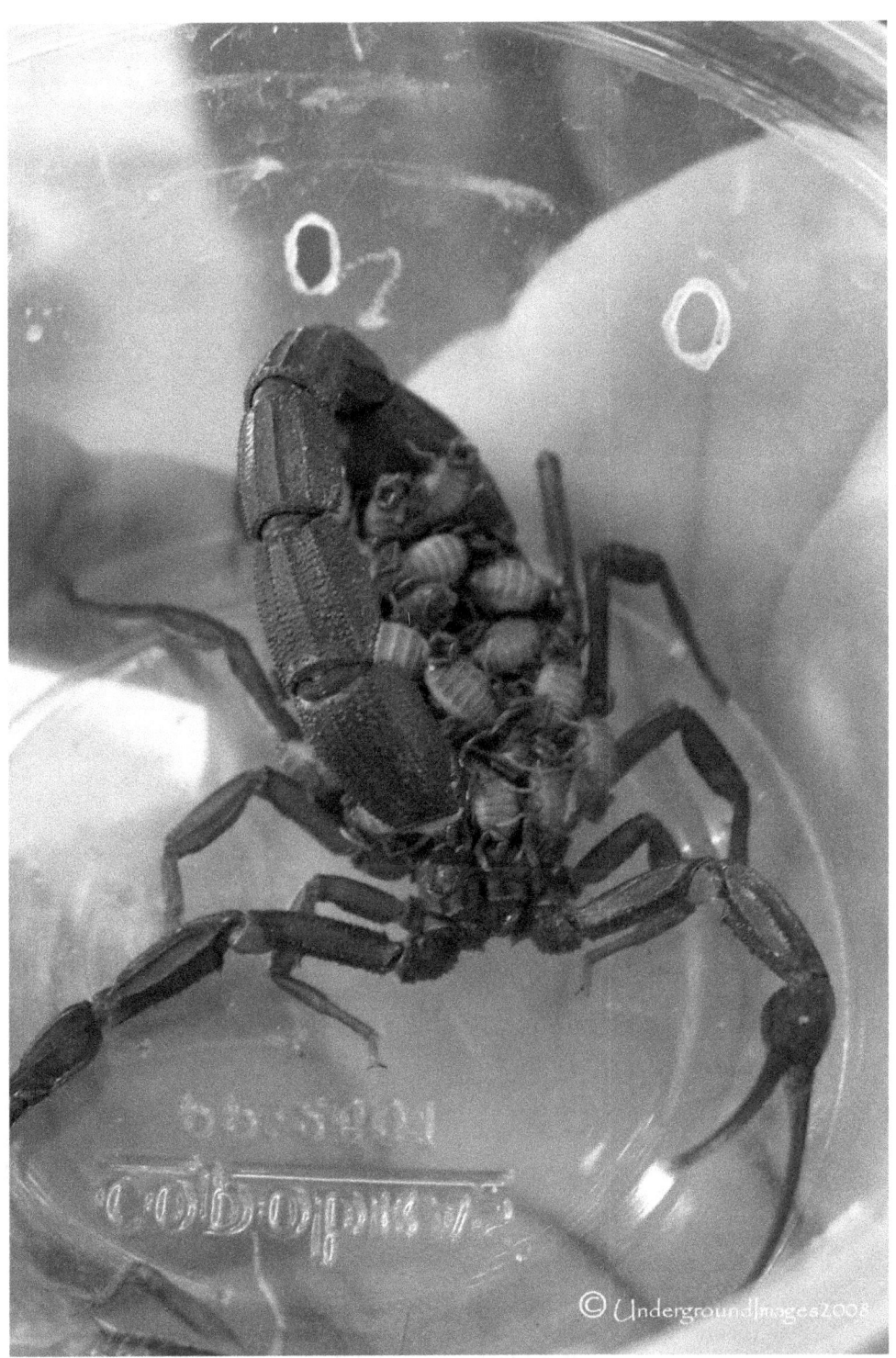

The new species of scorpion

Two flights - **(above)** the silver river snakes through the rainforest **(below)** Biffo's fear of flying was increased when the pilot's seat collapsed into his lap mid-flight

Above: Richard and Jon with young friends
Below: Giant river otter *(Pteronura brasiliensis)* kept in disgusting conditions at Georgetown Zoo

Above: Richard `phones home
Below: Joseph points out the mountain where the `plane crashed

A caboura fly feasts on Damon's blood

Above: Tea with Ernest's family
Below: The Chase Vault famous for the creeping coffins of Barbados

GIANT SNAKES

It is in South America that we meet with the most numerous reports of outsized ophidians. This is unsurprising, as the neo-tropics is the lair of the giant snake *el supremo* - the anaconda. In terms of bulk, this snake is by far the largest in the world. It's girth far greater than that of the reticulated python. Ever since the white man first ventured tentatively into the `green hell`, he has brought back tales that are the very stuff of nightmares - snakes whose size defies belief.

The earliest man to return with such bone chilling yarns was one Charles Waterton (1782-1865) - better known as Squire Waterton - a great British eccentric and adventurer. A Yorkshireman from a wealthy Roman Catholic family, The Squire insisted on sleeping on bare-boards with a block of wood as his pillow. Almost unique in his age, he was a teetotaller and violently opposed to hunting for sport. He was a passionate naturalist, and collector of animals, and with true intrepid Yorkshire spirit, he made four expeditions to South America between 1812 and 1824 - travelling in Brazil, Venezuela, and Guiana.

In typical Waterton style, he exposed as much of his skin as he could in the jungle at night, hoping to be bitten by a vampire bat. He was most disappointed when he was not bitten - but one of his companions was. The ungrateful man ran and hid in a latrine. Waterton`s books are full of such shenanigans, and it is obvious he enjoyed himself immensely. The Squire lived to the ripe old age of 83, a miracle when one reads of some of the risks he took!

Of the anaconda he writes:

"The camoudi snake (as it was called in British Guiana) has been killed from thirty to forty feet long; though not venomous, his size renders him destructive to the passing animals. The Spaniards in the Oroonoque positively affirm that he grows to the length of seventy or eighty feet and that he will destroy the strongest and largest bull. His name seems to confirm this; there he is called "matatoro" which means literally "bull killer". Thus he must be ranked among the deadly snakes, for it comes to the same

thing in the end whether the victim dies by poison from the fangs, which corrupts his blood and makes it stink horribly, or whether his body be crushed to a mummy and swallowed by this hideous beast".

A missionary Farther de Vernazza wrote in the 19th century surely what is the most fatuous description of the anaconda:

"The sight alone of this monster confounds, intimidates and infuses respect into the hart of the boldest man. He never seeks or follows the victim upon whom he feeds, but so great is the force of his inspiration, that he draws in with his breath whatever quadruped or bird may pass him within twenty to fifty yards of distance, according to its size. That which I killed from my canoe upon the Pastaza (with five shots from a fowling piece) had two yards of thickness and fifteen yards of lengths; but the Indians have assured me there are animals of this kind here of three or four yards in diameter, and from thirty to forty long. These swallow entire hogs, stags, tigers, and men, with the greatest facility".

The good father was confusing diameter with circumference methinks, else his snakes would be extremely stout. Alternatively, he may have shot one that had just eaten a large prey-item such as a tapir. The super snake suction he speaks of is total fantasy, but the supposed mystic effects of anaconda breath is a stubborn myth - as we shall see in a moment.

Another 19th century yarn that stretches the imagination, was told by a botanist called (appropriately) Dr Gardner. Whilst travelling in the province of Goias - near the head waters of the Araguaia River - his host's favourite horse disappeared from its pasture, and could not be found despite an intensive search. Finally they came upon the bloated body of a giant snake in a tree. It was dragged out into the open by two horses and found to measure eleven metres (thirty-seven feet) in length. When slit open, it was found to contain the broken hal-digested bones of the missing horse. The unfortunate animal's head was intact.

In fact, the anaconda is an aquatic rather than arboreal snake. Its great weight makes it a poor climber when adult. This would have been compounded by such a heavy meal. making its treetop siesta an impossibility. Also, a full grown horse would be exceptionally hard to swallow even for a thirty-seven foot snake. As well as having a large body, a horse possesses very long legs that cannot be readily folded back against its trunk. It would seem that thirty-seven feet would be the bare minimum length that a snake would need to be, to achieve such a feat.

In 1944 another specimen of this size was encountered in Columbia by a team of prospecting geologists led by Roberto Lamon. The men shot the snake and measured it at 11.4 metres (37 feet 6 inches). The group left the creature to eat their lunch, intending to come back and photograph their trophy and skin it. Upon their return they were amazed to find it gone. The bullets had merely stunned the animal which had recovered and absconded in their absence.

Fredrico Medem - a Columbian herpetologist - saw an anaconda that he estimated to be between nine and twelve metres (thirty-forty feet), and obtained a report of another thirty-four feet long.

General Candido Mariano de Silva Rondon - who lent his name to the Rondonia area of Brazil - saw a specimen killed by Indians, some 11.6 metres (thirty-eight feet) long. There are several records of snakes in this size bracket that cannot easily be dismissed, as some have involved reputable scientists. A 10.4 metre (thirty-four foot) anaconda was shot by Vincent Roth, director of The National Museum, in British Guiana (now Guyana). Mr R. Mole - a naturalist who made many important contributions to the knowledge of the wildlife of Trinidad - reported a ten metre (thirty-three foot) example there in 1924. Dr F. Medem of the Colombia University, saw a 10.26 metre (33 foot 8 inch) snake killed on the Guaviare River.

In 1909 war was on the verge of exploding in South America. A "rubber rush" to rival the gold-rushes of the old-west was happening, and a dispute was occurring in the Rio Abuna rubber plantations on the western borders of Brazil. Peru and Bolivia also meet at this point, and a bitter wrangle between the three countries over the valuable resource was growing to dangerous levels. Into this drama, The Royal Geographical Society sent a mediator to defuse the situation. Major Percy Fawcett - a 39 year old artillery officer - was to make the first intensive study of the area.

It was whilst engaged in this task that he initially herd of giant snakes. The manager of a remote hamlet called Yorongas, told him that he had killed a fifty-eight foot anaconda in the lower Amazon. Fawcett disregarded the story at first, but subsequently claimed to have shot an even bigger specimen.

Several months after the conversation at Yorongas he was on the Rio Abuna, upstream from it's junction with the Rio Rapirrao when:

"....almost under the bow of the igarite there appeared a triangular head and several feet of undulating body. It was a giant anaconda. I sprang for my rifle as the creature began to make it's way up the bank, and hardly waiting to aim smashed a .44 soft-nosed bullet into it's spine 10 feet below the wicked head. At once there was a flurry of foam, and several heavy thumps against the boat's keel, shaking us as though we

had run on a snag. With great difficulty I persuaded the Indian crew to turn in shoreward. They were so frightened that the whites showed all round their popping eyes, and in the moment of firing I had heard their terrified voices begging me not to shoot lest the monster destroy the boat and kill everyone on board, for not only do these creatures attack boats when injured, but there is also a great danger from their mates.

We stepped ashore and approached the reptile with caution. It was out of action, but shivers ran up and down the body like puffs of wind on a mountain tarn.. As far as it was possible to measure, a length of 45 feet lay out of the water, and 17 feet in it, making a total length of 62 feet. It's body was not thick for such a colossal length-not more than 12 inches in diameter- but it had probably been long without food. I tried to cut a piece out of the skin, but the beast was by no means dead and it's sudden upheavals rather scared us. A penetrating foetid odour emanated from the snake, probably it's breath, which is believed to have a stupefying effect, first attracting then paralysing it's prey. Everything about this snake was repulsive.

Such large specimens as this may not be common, but trails in the swamps reach a width of 6 feet and support the statements of Indians and rubber pickers that the anaconda sometimes reaches an incredible size, altogether dwarfing that shot by me. The Brazilian Boundary Commission told me of one killed in the Rio Paraguay exceeding 80 feet in length."

This is the most celebrated and oft repeated encounter with a giant anaconda, but it is also one of the most questionable.

- Firstly, the width given for this snake is absurdly small. The anaconda is a massively built snake. A specimen *half* this length would have a width twice as wide or more. Fawcett's snake would have had to be an emaciated near-skeleton!
- Secondly his assertion that "there is a great danger from their mates", implies that anacondas mate for life and their partners will seek revenge for the killing of a mate. This is nonsense, no snakes are life-maters, and anacondas breed in huge "mating balls". These consist of dozens of males competing to mate with one larger female.
- Finally no snakes have "stupefying breath". This idea of breath that draws in and paralyses prey can be traced back to dragonlore. The breath of a giant anaconda may well be foul but it possesses none of these attributes.

For these reasons, I am inclined to reject Fawcett's story as a traveller's tale. The man himself disappeared several years later whilst looking for a lost city in the jungle.

There are other accounts however, which are not so easily dismissed, and the anaconda has one huge advantage over the python that may well allow it to attain a greater size. All pythons are oviparous - that is they lay eggs. This must be done on land. Anacondas are ovo- viviparous - they retain the eggs *inside* their bodies until the young hatch, then give birth to them live. This means they do not have to leave the water - their final link with the land is broken. Living in water almost all of the time, means anacondas are buoyed up - they do not have to support their own body weight on land very often, and hence can grow to a very large size.

The Marquis de Wavrin was another explorer of South America, and was active in the years before the Second World War. He told the great Belgian Cryptozoologist Bernard Heuvelmans, that he had seen anacondas over thirty feet long, and that the natives told of far larger ones. He once shot an eight metre (twenty-six foot), individual that had been coiled around a branch. When he expressed a desire to retrieve the cadaver, his canoe-men told him that it was a waste of powder to shoot such a small snake and a waste of time picking it up. They went on to say:

"On the Rio Guaviare, during floods, chiefly in certain lagoons in the neighbourhood, and even near the confluence of this stream, we often see snakes that are more than double the size of the one you have just shot. They are often thicker than our canoe."

F.W Up de Graff - an explorer of seven years experience - spotted a giant anaconda as it lay in shallow water under his canoe:

"It measured fifty feet for certainty, and probably nearer sixty. I know this from the position in which it lay. Our canoe was a twenty-four footer; the snake's head was ten or twelve feet beyond the bow; it's tail a good four feet beyond the stern; its body was looped into a huge `S`, whose length was the length of our dugout and whose breadth was a good five feet."

Algot Lange claims to have shot a seventeen metre (fifty-seven foot) anaconda, and skinned it. Willard Price, the author of the wild life "adventure" series (wherein youthful heroes travel the world capturing wild animals for zoos in unlikely adventures, that in zoological terms, are about as accurate as a lobotomised chimp with Parkinson's disease attempting to forge a Rembrant!), says he then took the hide to New York. He also claims the snake "mesmerised" his men (snake's hypnotic powers are a myth). Of course he - as the white man - had enough mental-power to resist the snake's trance inducing gaze and save the day by machine-gunning the offending reptile. Nothing more was heard of the hide and given Price's love of bullshit, it is perhaps best to write this off as a hoax.

Another tall tale that involved a hypnotic anaconda was related by Harold.T.Wilkins in his 1952 book *Secret Cities of Old South America*. Wilkins heard the tale from one Alfred.G.Hales, who had - in turn - gleaned it from Indians far up the Brazilian Amazon. An Indian was fishing one night in his canoe, when two moon-like lights on the bank, attracted him, and drew him to the shore. The lights were - in fact - the eyes of an anaconda. Just as he was about to reach the shore, another giant reptile (hinted at as being a dinosaur), lunged up from the river and snatched the giant snake from the branches.

The natives had also told Hales that anacondas lurked near temples, luring parrots from the trees with their hypnotic eyes that changed colour from green to red!

When witnesses are cross examined - face to face - by a renowned zoologist, we have to give them a little more credence. One of the witnesses of the next case was interviewed over several days by no less an authority than Heuvelmans himself.

It was in 1947, when many wild and sometimes fierce tribes were still commonplace in the South America. A particularly warlike group were The Chavantes - who had recently killed a number of Brazilian officials. Francisco Meirelles of the Service for the Protection of the Indians, organised an expedition to try to establish peaceful relations with this tribe. The five month endeavour included in its twenty-man line up. Serge Bonacase - a French painter whom Heuvelmans later interviewed.

By the second month, the company had reached a large island between the two branches of the Araguaya river, and made base-camp there. The men spent several days in preparation for the big push into the wilderness (or the *sertao* as the `green hell` was known). They spent long reconnaissance and hunting trips away from the island. On one such trip, eight of them were hunting capybaras in a swamp between the Rio Manso, (charmingly known as the *Rio das Mortes* - `The River of Death`, as The Chavantes butchered any one who dared to cross it), and the Rio Cristalino. The Chavantes did not put in an appearance, but the group encountered something far more frightening.

"The guide pointed out an anaconda on a rise in the ground half hidden among the grass. We approached to within 20 metres of it and fired our rifles at it several times. It tried to make off, all in convolutions, but we caught up with it after 20 or 30 metres and finished it off. Only then did we realise how enormous it was; when we walked the along the whole length of it's body it seemed as if it would never end. What struck me most was it's enormous head.

As we had no measuring instruments, one of us took a piece of string and held it be-

tween the ends of the fingers of one hand and the other shoulder to mark of a length of one metre. Actually it could have been a little less. We measured the snake several times with this piece and always made it 24 or 25 times as long as the string. The reptile must therefore have been nearly 23 metres long."

Unfortunately, none of the men were zoologists, and none realised the importance of the find. Bonacase himself had heard so many stories of giant anacondas he believed them to be commonplace. The carcass, and even the skin, would have weighed the men down too much for them to have brought it back. So, sadly, this invaluable specimen was left to the jungle scavengers. (This seems to be the bane of cryptozoologists. Specimens always fall into the hands of those who do not know their importance, and hence seldom find their way to civilisation.)

The late 1950s brought perhaps the most dramatic encounter with an anaconda. The political climate, with its upsurgence in communism in Latin America, was such that the U.S government placed CIA agents in sensitive areas. One agent - called "Lee" - was told by a cattle-rancher of a giant-snake lairing in a cave in Bolivia. The reptile was said to be over ten metres (thirty-three feet) long. It was said to have eaten ten Indians and many cattle over the years. Every three months, or so the serpent emerged, seized a steer, dragged it into the river, killed it, then ate it. Then it would return to its cave.

The rancher wanted Lee to capture the animal and take it too a zoo as it was "probably the largest snake in the world". The problem was discussed at The Embassy many times until someone came up with an audacious plot to catch it. The plan was to flush the monster from its lair with tear-gas whilst a long sack (complete with zip fasteners), was held over the caves mouth. There would be two "zip-men" - one at each end of the sack - to hasten the operation. For added security Lee carried (ironically) a .357 python pistol.

It was just as well Lee was "packing heat" as things did go spectacularly wrong. The tear-gas was shot into the cave, and the anaconda - thrashing madly - shot out of the cave, and into the sack. Once its entire length was inside, both ends were zipped up. The agents had not reckoned with the snake's vast strength however. Its violent writhing split the sack - end to end - and the brute was free.

The livid animal came rushing at Lee who whipped out his pistol, and managed to put a bullet in its head. The snake threw itself into a huge loop, smashing into a small hardwood tree about as big as a telephone pole. The tree was shattered like matchwood and the snake fell back into the jungle. Lee pumped another two bullets into its head. When it had expired they measured it Its length proved to be thirty-four feet three inches. Lee skinned the snake and took the hide back to the United States where he kept it in his garage. Its current whereabouts are unknown. As noted earlier, this size would seem very small for a snake which was able to swallow such large livestock.

Lee's colleague David Atlee Phillips, understandably doubted his friend's outlandish story. Sometime later he was attending a party at Washington, and mentioned the saga to Darwin Bell, then Deputy Assistant Secretary for International Labour Affairs. Bell claimed not only to have known Lee but to have taken part in the capture attempt. *"I was the tail zipper man"*, he told an amazed Phillips.

More recently, a giant anaconda was reported near Sao Paulo Brazil. Farmer-come-hunter Joao Menezes was fishing with his three year old son Daniel, and turned his back on the boy to store some fish in a wooden shack. Suddenly his son's screams rent the air, and the horrified Menezes, turned to see a foprty-five foot anaconda had risen from the waters and sized his boy by the neck. He tried in vain to prise the snake's jaws apart then ran home for his rifle. By the time he got back, however, the boy had been crushed and was in the process of being swallowed.

More recently still, the world renowned explorer Colonel John Blashford-Snell was told a most intriguing story whilst travelling across the Andes by river from Bolivia to Bunenos. It seems that a thirteen metre (forty-three foot) anaconda was captured by a farmer after it had eaten a cow. He apparently incited it with a pig on a rope. Subsequently he tried to sell his story, unsuccessfully, to the press. The creature is now said to be residing in a pond on a farm in north west Brazil. This occurred in late 1999.

If the creature is being fed by the farmers it may well remain in the pool. This is a cryptozoological "sitting duck". If the story is true is should be child's play to find and film this giant. The author hopes to find financial backing to do just that!

Even 45-66 footers seem like runts in comparison to some of the claimed monsters. There is a school of thought that there are two separate species of giant constrictor in the Neo-tropics, the giant anaconda and the markedly different and far larger *sucuriju gigante* or giant boa.

The Marquis de Wavrin - whom we met earlier - was told of such behemoths by his canoe-men who seemed to think them different to anacondas. Once, when the Rio Uva was in flood, some Piapoco Indians tried to take a short cut to the Rio Guaviare via some marshes and lagoons. Having just crossed a small lake, the Indians heard a sound akin to thunder behind them - even though the rains had ceased and the skies were clear. Looking back, they saw the waters in turmoil, as a massive animal thrashed about in mid-water. Then a gigantic snake's head broke the surface, and the animal disported itself momentarily, before diving again. (We should note here the interesting parallels with Oriental Dragons, and their association with rain and storms). The Indians believed that had the monster surfaced whilst they where crossing, they would have been devoured. Not unreasonably, they vowed never to take that particular short cut again.

The Marquis himself only narrowly missed seeing a *sucuriju gigante*. He reached the Rio Putumayo the day after a giant boa had dragged off an ox. The people were still in a state of shock. He writes of these giants:

"Around the upper Paraguay they give the name minocao to a more or less fabulous snake: the natives say it can reach the size of a canoe. They suppose that it is a sucurijiu or a boa-constrictor that has grown very old and turned into a water snake. On the upper Rio Parana, in Brazilian territory, I have also been told of these enormous snakes, capable of dragging a canoe to the bottom. These monsters frequent deserted places, and never leave a river. The fear they arouse is quite superstitious."

This idea of a giant serpent growing from as small snake, is also seen in Asian dragon legends. The concept of a snake becoming too large to live on land and hence taking up an aquatic lifestyle, echoes the Scandinavian Lindorm stories.

One man who was convinced of the giant boa's existence, was Lorenz Hagenbeck - director of Hamburg Zoo and son of the famous Carl Hagenbeck. The Hagenbecks were a dynasty of animal collectors who had supplied zoos worldwide with animals for over a century. In the days when captive breeding programmes were only a twinkle in zoo curator's eyes, the Hagenbecks provided rare and exotic beasts from all around the globe. It was one of Hagenbeck's explorers who first discovered the pigmy hippopotamus, (*Choeropisis liberiensis)*, in Liberia on 28th February 1913. Lorenz himself discovered the skin of the Andean wolf, (*Dasycon haganbecki)* in a Buenos Aires market in 1927. As far as we know, no westerner has ever seen this animal alive and only one skull from this rare animal has fallen into the hands of zoologists.

The Hagenbeck family papers include several reports and transcripts pertaining to the "mother of all snakes".

Two of Hagenbeck's confidantes were Roman Catholic priests, Father Victor Heinz and Father Protesius Frickel. Father Heinz was lucky enough to see the giant boa on more than one occasion:

"During the great floods of 1922 on May 22 - at about three o'clock to be exact - I was being taken home by canoe on the Amazon from Obidos; suddenly I noticed something surprising in midstream. I distinctly recognised a giant water snake at a distance of some thirty yards. To distinguish it from the sucurijiu, the natives who accompanied me named the reptile, because of its enormous size, sucurijiu gigante (giant boa).

Coiled up in two rings the monster drifted quietly and gently downstream. My quaking crew had stopped paddling. Thunderstruck, we all stared at the frightful beast. I reckoned that its body was as thick as an oil drum and that its visible length was some eighty feet. When we where far enough away and my boatmen dared to speak again they said the monster would have crushed us like a box of matches if it had not previously consumed several large capybaras."

Spurred on by such a dramatic encounter, the priest began to study the phenomenon seriously. He discovered that another specimen had been killed, a day's march from Obidos, as it was in the act of swallowing a capybara (*Hydrochoerus hydrochaeris*), the world's largest rodent. This semi-aquatic species resembles a giant guinea pig and grows to the size of a large dog and are one of the principle prey-species of the anaconda. This particular snake had been on the shore of Lago Grande do Salea. Its stomach contained four adult capybaras. Elsewhere, two huge round scats were discovered and attributed to the giant boa. They contained animal hair. One had an oxen's hoof bone protruding out of it. The snake itself was far from finished with Farther Heinz.

"My second encounter with a giant water snake took place on 29 October 1929. To escape the great heat I had decided to go down river at about 7 p.m. in the direction of Alemquer. At about midnight, we found ourselves above the mouth of the Piaba when my crew, sized with a sudden fear, began to row hard towards the shore. "What is it?" I cried, sitting up. "There is a big animal", they muttered very excited. At the same moment I heard the water move as if a steamboat had passed. I immediately noticed several metres above the surface of the water two bluish-green lights like the navigation lights on the bridge of a riverboat, and shouted: "No, look, it's the steamer! Row to the side so that it doesn't upset us." "Que vapor que nada", they replied. "Una cobra grande!" Petrified, we all watched the monster approach; it avoided us and recrossed the river in less than a minute a crossing that would have taken us ten to fifteen times as long. On the safety of dry land we took courage and shouted to attract the attention of the snake. At this very moment a human figure began to wave an oil-lamp on the other shore, thinking no doubt, that someone was in danger. Almost at once the snake rose on the surface and we were able to appreciate clearly the difference between the light of the lamp and the phosphorescent light of the monster's eyes. Later, on my return, the inhabitants of this place assured me that

above the mouth of the Piaba there dwelt a sucuriju gigante".

This account is interesting from a zoological point of view, because it contains a detail unknown to Father Heinz. The priest and his friends tried to attract the snake by shouting, to no avail. However the animal responded to the light stimulus. This is because snakes are deaf - a fact Heinz and co were clearly unaware of. This lends the report some weight and as far as I know has not been commented on before.

Heinz began to interview other witnesses. One of these was Reymondo Zima, a Portuguese merchant who had lived for nine years opposite the town of Faro on the Rio Jamunda

"On 6th July 1930 I was going up the Jamunda in company with my wife and the boy who looks after my motor-boat. Night was falling when we saw a light on the river bank In the belief it was the house I was looking for I steered towards the light and switched on my searchlight. But then we noticed that the light was charging towards us at an incredible speed. A huge wave lifted the bow of the boat and almost made it capsize. My wife screamed in terror. At the same moment we made out the shape of a giant snake rising out of the water and performing a St Vitus`s dance around the boat. After which the monster crossed this tributary of the Amazon about half a kilometre wide at fabulous speed, leaving a huge wake, larger than any of the steamboats make at full speed. The waves hit our 13 metre boat with such force that at every moment we were in danger of capsizing. I opened my motor flat out and made for dry land. Owing to the understandable excitement at the time it was not possible for me to reckon the monster's length. I presume that as a result of a wound the animal lost one eye, since I saw only one light. I think the giant snake must have mistaken our searchlight for the eye of one of his fellow snakes.

In the same area, in 1948, an old pupil of Farther Heinz - Paul Tarvalho - had a sighting of his own. He observed - from a distance of some 900 feet - a gargantuan snake emerge from the water. Tarvalho estimated it to be fully fifty metres (167 feet) long! This mammoth serpent followed his boat for a moment, and needless to say Tarvalho made off a top speed.

Father Frickel was brave (or foolish) enough to approach one of these titans on land. Whilst on an expedition on the upper reaches of the Rio Trombetas, he saw the head of a giant boa lying in water by the bank. The foolhardy father approached to within six paces of it, and noted its eyes were as large as dinner plates. This is another difference to the true anaconda whose eyes are comparatively small and beady.

As a one time zookeeper specialising in reptiles I can vouch for the astounding strength of constricting snakes. Once, a young Indian python no more than three feet long lifted the Perspex lid of its tank and escaped for several days. The lid had been weighed down with several large rocks, but these had proved no obstacle to the snake. One can imagine the titanic strength wielded by a sucuriju gigante that - if reports are to be believed - can exceed 150 feet in length! Father Heinz had a story related to him demonstrating just how powerful the giant boa is.

"On 27 September 1930, on an arm of water that leads from Lake Maruricana to the Rio Iguarape, a Brazilian named Joao Penha was engaged in clearing the bank to make it easier for turtles to come up and lay their eggs. At a certain moment, behind one of those floating barriers made of plants, tree trunks, and tangled branches, against which steamers of 500 tons often have to battle to force a passage, he saw two green lights.

Penha thought at first that it was some fisherman who was looking for eggs. But the whole barrier shook for 100 metres. He had to retreat hurriedly from a foaming wave 2 metres high struck the bank. Then he called his two sons, and all three of them saw a snake rising out of the water pushing the barrier in front of it for a distance of some 300 metres until the narrow arm of the water was finally freed of it.

During all this time they could observe at leisure its phosphorescent eyes and the huge teeth of its lower jaw."

Three photographs exist - apparently showing specimens of the giant boas killed in 1933, 1948 and 1949. The first two were published in Rio de Janeiro newspapers. All of these had been developed by the same man - Miguel Gastao, the proprietor of a bazaar at Manos. Farther Heinz interviewed him, and was assured that none of the photos had been tampered with.

The first had been brought in by the Brazilian-Colombian Boundary Commission, who said the snake been killed on the banks of the Rio Negro. The thirty metre (100 foot) creature had been machine gunned and in its death-throes had reared up nine metres (thirtyfeet) crushing bushes and small trees under its two ton bulk. Four men had been unable to lift its head.

The second was captured alive whilst swallowing a steer (the bull's horns were still protruding from the snake's mouth). A rope was affixed about its neck and it was towed into Manos (this begs the question why such a large and powerful animal behaved in such a placid way). There it was finally killed by machine-gun fire. It was alleged to be forty metres (131 feet) long and to weigh five tons!

Tim Dinsdale, the late monster hunter and world renowned Cryptozoologist, examined the photo published in the *Diario de Pernambuco* on January 24 1948. He said:

"One of the first things that struck me looking at this photograph, was that although the beast appeared to have an enormous blunt snouted head of a giant snake type it was altogether different from the anaconda, boa constrictor, and python. For one thing it had eyes that were much too large, and a great bag of a mouth, the mottled white markings on which where unlike those on the anaconda, which in old age sometimes develops jowls beneath the lower jaw......

....Another thing: in the photograph at the sixth convolution the body is at its visible greatest. Whereas on the largest of known snakes, the reticulate python, skeletal rib-structure is clearly greatest at the fourth convolution."

I have never actually seen a copy of this picture, but Dinsdale`s observations seem to be born out by his detailed sketch taken from the original.

The second 1948 photograph was taken after a dramatic encounter in the ruins of Fort Tabatinga on the River Oiapoc in Guapore territory. The thirty-five metre (117 foot) snake crawled ashore and made its lair in the fort, (just like a medieval dragon). It took an amazing five-hundred rounds of machine gun fire to kill it. The titanic cadaver was photographed as it floated down river. The picture shows a large snake floating belly up in water. The far bank is visible with buildings on it, but without knowing the width of the river and how far the body is from the bank, we cannot estimate its size. The extremities are below the water and the body seems distended with gas. I have only seen black and white prints of this, but Dinsdale - who saw a colour copy - said its belly was a mottled-white, a description that does not fit the anaconda.

Naturalist Peter Matthiessen, was told by Fausto Lopez - a hotelier from Pucallapa, Peru - that in the 1950s he saw an old and harmless "anaconda" thirty metres (100 feet) long, that had killed by Indians on the Huallaga river. How any snake measuring 100 feet can be considered harmless is quite beyond the comprehension of this author.

So far, all the sightings of the giant boa have been in South America, but there is one report that suggests they might range further north. Francois Poli was a French explorer who studied shark fishing in Lake Nicaragua. In his book *Sharks are Caught at Night*, he recounts how he met a German called Brennecker, who encountered a giant boa on the borderlands of Nicaragua and Honduras:

"I was driving a jeep along a sort of natural track winding between two lines of trees when I saw, about 50 yards ahead, a huge fallen tree trunk which barred the way. I

told the boy who was with me to find some way of shifting it. He came back at a run. It wasn't a tree-trunk at all, but a snake. It stirred and began moving slowly towards us...

I've seen the most incredible snakes in this country during the past twenty years- and I can assure you I know how to handle a gun. But that day I left the revolver where it was; I just stepped on the gas and drove off."

Jeremy Wade interviewed several Amazonian fishermen, whilst he was exploring the area in 1995. His first informant was Dorgival Sabino, who saw a giant boa on the Rio Negro (seemingly the place for them):

"It was a gigantic animal, like a monster. A snake, but of a size much bigger than normal with the difference that its head was like some kind of dinosaur, with- I don't know whether they were teeth or horns, just that it was grotesque."

Sabino put its length at twenty metres (sixty-five feet), and its width at a metre (three feet). The horns are more problematical. The only horn-bearing snakes are all vipers (such as the rhinoceros viper and the horned viper). It has been suggested that the horns, that have been reported on giant snakes several times, are in fact the horns of bulls that the snake is swallowing protruding from the corners of its mouth.

On a tributary called the *Rio Purus*, Wade met another witness - Amarilho Vincent de Oliveira. Whilst navigating the backwaters one night some twenty years before, Oliveira and a companion came upon what looked like a floating-tree in his torch light. As they passed it, he looked back. and saw the "tree" had turned ninety degrees to face them despite there being no current. They doubled back and crossed the ninety-metre channel paddling stealthily up the other side. The "tree" again turned to face them. This time they got a look at it, and it became apparent that this was no fallen tree.

"Its head had horns like the roots of a tree, and could see these greenish eyes as well. We just left the canoe on the bank and got out of there. Afterwards, people saw movements in the water there, many times. With no wind, there would be waves that covered the beach."

Again, I think that a prosaic explanation for the horns can be found. You will recall Joao Penha`s description of a giant boa pushing its way through a tangled mass of vegetation that was blocking the river. Could not this specimen have done the same and still retained some branches caught on its head? By torch-light these could have been mistaken for horns.

In 1996 Wade returned to continue his inquires. This time he travelled to an even more remote area. One 76 year old man described how as a boy he saw a rib cut from a giant snake that had been shot, but which was too large to drag ashore. The rib was a metre (three feet) long. The snake was killed by an engineer who had spotted a strange light on the river just before a large wave washed his boat onto a bank.

Botanist Grace Rebelo dos Santos, told Wade that in June 1995, she saw two lights appear in the middle of the river. The lights emanated from a place that earlier that evening, a dragnet had become caught on something very heavy that had then escaped.

"It came right in close to the bank then disappeared. The lights were like torches, about 30 cm apart.. I'm not going to say it was a cobra-grande but I remember clearly how blue the lights were, which I thought very strange"

What are we to make of these claims? Do snakes in access of 150 feet actually exist? The answer is that currently no one knows for sure. Such monsters seem fantastic, but so did the giant squid, the gorilla, and the okapi before they fell into the hands of western science. We have seen in the previous chapter, that there is a fossil precedent for giant snakes in South America. Perhaps these antediluvian constrictors are not extinct at all, but still linger in the dim, steamy, interior of the Amazonias.

Could these gigantic serpents be of a prehistoric lineage thought long extinct? One group of fossil snakes that I have mentioned before - the *Madtsoids* - did reach huge sizes. A fossil rib from South America suggests a snake of eighteen metres (sixty feet). The *Madtsoids* were at first thought to be giant boas or pythons, but we now know they belong to a more primitive basal group of snakes. Despite their primitive nature, they were highly successful and spread across the world. Evolving in the *Cretaceous* period - over 100 million years ago - some, such as the Australian *Wonambi*, lingered until only 10,000 years ago. Could some species have survived and grown to even more titanic proportions in remote jungle areas of the world?

Edited from:
Dragons:More than a Myth?
by Richard Freeman,
CFZ Press, 2005

KEEPING THE HOME FIRES BURNING

People often come up to me and say that I must have a wonderfully romantic life, because I am a professional monster hunter. Well, it is true, but it is a particularly peculiar existence. Much of my time is spent in boring administrative duties, and as the CFZ gets larger, more corporate, and therefore more complicated to run, I cannot see my life becoming any easier, in the short term at least!

People often also ask me whether I mind being left behind in the UK, whilst the younger and fitter members of the expedition go gallivanting off to foreign climes. It is true that I haven't been on a foreign trip since November 2004, but that has mostly been because of personal reasons, and my wife Corinna and I have some foreign expeditions planned for the near future. But, no - I don't mind at all. The CFZ is a team effort and is far more than the sum of its component parts, and - especially in the new era of global communication - I am in the enviable position of being able to coordinate what is happening on the other side of the world from my own office.

The expedition had a SatPhone, and kept in touch most days, and my job was to translate the patchy - often almost incoherent - transmissions into some degree of understandable English, and then to post the news on the website.

During the weeks and months prior to the expedition I did my best to drum up enthusiasm amongst the inhabitants of cyberspace, and two days before the expedition was to embark on the 14th November, I wrote:

"Well guys we are nearly there. On Wednesday evening the intrepid team of four men and one woman of the CFZ embark for Guyana. Paul Rose (God bless him) has obtained a satellite 'phone, and – technology permitting – we hope that we will be able to post daily updates on this blog. You will have to wait until they return on the 28th November before we can post any pictures or video, but we are confident that our daily bulletins from the expedition will be an exciting opportunity for you all to follow the adventure as it unfolds."

I had a brief telephone call from Richard at tea-time on the 14th, telling me that all was well, and that the team had assembled at the airport, and then…..

… as Paul Rose would no doubt have said, a big fat NOTHING!

Nearly 24 hours later, I received a fifteen second telephone call from Richard in Guyana. My problem was that in the intervening time, I had been inundated by telephone calls and eMails from the media, from concerned well-wishers, and from friends and family of the expedition members asking for news. There wasn't any, but in the best traditions of Her Majesty's \Press, I released the following statement:

> *""the good news you cannot refuse,*
> *the bad news, is there is no news ..."*
> Scott Walker *Tilt* 1995

"This is the fifth time in as many years that I have spent at home in Britain while Richard Freeman and Chris Clark lead a major foreign cryptozoological expedition. I do not mind doing it, because I am only too aware that each expedition involves physical rigours with which my poor battered middle-aged body is no longer able to deal. As many of the readers of this blog know, I am seriously disabled, and although I have been on foreign expeditions in the past, and no doubt shall do so again, sixteen-hour treks into the jungle are sadly a thing of the past for me. However, in my opinion, my job; staying at home and collating the information that we receive is just as important as being on the front line. No army marches without its logistical support from GCHQ, and the CFZ expeditionary force is no exception.

The way I see it, is that the CFZ in funded by public subscription, and therefore, you - the public – deserve to be kept up to date with the progress of each of our expeditions. We have every intention of posting the news as and when it happens. However, sadly, the world's an imperfect place, and sometimes technology does not work as well as we would have hoped.

I am particularly unwell at the moment, and have been in bed most of the day. Indeed, it is only through the kind offices of my wife Corinna, who is sitting typing this as I dictate it, that the news bulletin you are reading now is being typed.

We know that the Caribbean Airlines flight arrived in Georgetown at lunchtime GMT, and we have been anxiously waiting for some word from the expedition team ever since 1.00 pm.

At about 6.30 tonight, we had a brief 'phone call from Richard. Oll Lewis was manning the office 'phone:

"Richard 'phoned and said: 'Hello is that Jon?' To which I replied 'No, it is Oll. I shall patch you up to Jon now', and the line went dead."

Oll has been trying to contact them ever since.

He says: *"I have got through to the 'phone twice, but both times British Telecom's automated bionic disembodied voice has cheerfully announced 'we are sorry, but there is a fault on the line'. The second time I got through it briefly announced that I was through to the 'phones voicemail, before the annoying voice of BT informed me of a fault again. I shall be trying to contact them again every 15 minutes."*

My old boss in the music industry, a very well-known pop star, once told me that one of my greatest talents was being able to concoct press releases out of nothing. Richard laughed when I told him this, and pointed out that I managed to get an eminently readable 120,000-word book after 4 days spent at the side of a lake in Lancashire. This bulletin, however, really takes the biscuit. We have had four words from Richard, from which we can assume that they are:

a) safely in Guyana, and
b) that none of them, as yet, has been eaten by giant ground sloths.
(I am waiting, with glee, for the first time that someone points out that ground sloths are/were exclusively vegetarians).

I promised you the news as and when we have it, and so far this is it. There will be more bulletins as and when more news is available.

Watch this space, "

The truth is that I was becoming worried. Not very worried, but worried enough to make a telephone to an expat Brit living in Dallas. Over in Texas, Nick Redfern was co-ordinating the transatlantic publicity campaign, and he, too was becoming a little concerned. We had not had enough information really to convince us that the team actually *were* alright, despite what we had told the world. I wondered whether, because at least Nick was in roughly the same time zone, that we might get better results if *he* telephoned the SatPhone…

Nope.

So he released the following statement on the CFZUSA blog:

"Well, the guys have arrived in Guyana and are deep within the depths of the jungles as I write these words. We heard from them earlier today - Richard (Freeman) called on his satellite phone to announce they were making their intrepid way towards one rumored beast-infested location...at which point the line went dead. Unfortunately - and somewhat mysteriously - they have not been heard of since. We sincerely hope they have not become lunch for some giant, marauding snake or hairy man-beast, and we anxiously await their next communication. Hopefully, this is all due to the effects of man-made satellite technology and nothing more sinister. When we hear more, so shall you..."

Whilst my earlier statement had kept most people happy, there were still some who were unconvinced, and so Nick tried - at intervals - to telephone Guyana from his Texas home.

Later that day, partly (I suspect) to soothe my nerves, and partly for comedic effect, he released another posting on his blog:

"The Guyana Five: Still M.I.A.

Well...the mystery of the missing adventurers (a.k.a. the "Guyana Five") continues at a steady and mystifying pace. A frantic, just-received phone call from CFZ director Jon Downes worriedly informed me that there is still no word from our intrepid explorers. Jon asked me if I would try and contact their satellite phone myself, given the fact that - here in Dallas - I'm much closer to Guyana than he is in ye olde England.

I was happy to oblige, of course; but had to ominously advise an ever more panicky and sweaty-palmed Jon that the only voice I heard was a recorded one: namely that of a posh English bird uttering the robotic words: *"There is a fault, please try later. There is a fault, please try later. There is..."*

Well, you get the picture. Should we now seriously consider the possibility that Richard and Co. really are nothing but a series of fragmentary and nostalgic memories to us, and the remains of a tasty morsel to the monstrous beasts of deepest Guyana? Probably not!

In reality, I'm quite sure that the expedition is continuing at a fine pace, and the gang is merrily roaming around, blissfully unaware of the fact that their fate - and their potential digestion by giant snakes - is being debated in slightly-more-than-half-serious words on this very blog.

Will we ever see the Guyana Five again?

Will they make it out of the lair of the beasts in one piece? Will Jon ever get to utter those immortal words again: "Richard, make me a nice cup of Earl Grey Tea, dear boy"?

Or will Jon be forever doomed to cry out late at night when the wind howls and the driving rain beats down on the windows of the ancient cottage in which he and his beloved Corinna live: *"Richard, Richard! Wherefore art thou, Richard?"*

There's only one way to find out: stay tuned for the next exciting and enthralling (well, okay, mildly intriguing) episode of what I like to call "Welcome to the Jungle."

When the Mission Controls both in the UK and the USA are truly concerned that something awful might have happened to the away team there is only one way to deal with it. With extreme silliness. T'is the CFZ way.

Then a few hours later we heard from the team properly, and were able to post the first *bona fide* news report:

"Less than a couple of hours after Nick Redfern wondered on his blog whether we would ever see our brave boys (and girl) again, we finally heard from Richard. OK the reception was terrible, there was five to ten second lag in broadcasting, and our conversation lasted for considerably less than two minutes, but we can no confirm that all five of the expedition members are safe and well, and that they are presently in the township of Letham Station on the edge of the savannah.

Despite the almost insurmountable technical difficulties, it still amazes me that I was able to have a telephone conversation with my good friend and colleague miles from nowhere, half the world away. For unlike the younger members of the CFZ team, I am a child of the late 50s, and I still remember how complicated it was to make my first ever international 'phone call. It was 1966, and my grandmother's 70th birthday. My father, who was then a senior civil servant in Hong Kong managed – somehow – to arrange an international 'phone call through somebody at the American Consulate so we could wish my grandmother a happy birthday. It took days to arrange, and I found out years later that my father had had to pull a lot of strings in high places to arrange it, and even now I remember the booming voice of my grandmother sounding like she had her head in a bucket along a crackly telephone line. Before we made the 'phone call my mother got out my father's globe, which still stands on top of my recording studio monitors in the CFZ office. She showed me the little dot that was Hong Kong, and the even smaller dot which was Chester in the north of England. I marvelled at this amazing technology, which could allow us to speak to my grandmother for a few minutes.

Some of this childish wonder came back to me this evening as I heard Richard's crackly voice broadcasting across the ether from thousands of miles away.

After ascertaining that everybody was well, Richard told us quickly of their first day's adventures. After an eighteen hour flight, they landed in Georgetown – the capital of Guyana – yesterday morning, and after a short break travelled for thirteen hours into the interior of the country. It's only when I checked Google Earth that I realised quite how far into the interior Letham Station actually is. The poor dears must be exhausted.

This morning they visited the home village of their guide and mentor Damon Corrie. The 'phone cut out for a few seconds as he was explaining this, so for the moment I cannot tell you the name of this village, but while they were there they received the first solid first person account of the didi.

Apparently, about two years ago, two children – a boy, and a girl aged twelve – were walking on the savannah near the village. Out of the undergrowth strode a big, hairy, man-like figure, who grabbed the little girl, disappeared with her, and neither the hairy didi nor the girl were ever seen again.

The annals of cryptozoology are full of such accounts, but it is a very disturbing feeling to hear such a story in the 21st Century. Cryptozoologists from Heuvelmans onwards have described the world's unknown hominids as shy and gentle creatures, and have done their best to downplay the stories of rapes, killings, and abductions. To hear such a story for oneself is a chilling experience.

They have also obtained the first video footage of an hitherto unknown species of scorpion, known to the locals as the green scorpion (presumably because of its colour). Richard was in the middle of telling me about this when the satellite link broke off. We were unable to contact him again.

All in all, however, this is a fantastic first day for the expedition.

Tomorrow the expedition sets out on foot into the largely unexplored savannah grassland.

Our thoughts and our prayers are with them."

The next day there was more news, and another new bulletin posted by us on the expedition blog, and then paraphrased by Nick for the US edition:

"Back at the CFZ headquarters in rural North Devon we are in a peculiar state of limbo. Our whole day's activities revolve about waiting for one garbled two or three minute 'phone call. After a long and fairly dull day, Richard 'phoned just before 10.00 this evening GMT.

It has been another long and eventful day for the five-person expedition. First of all we want to reassure you that everybody is alive and well, although they are finding the heat very difficult. After all, they are not that far from the Equator. It is a mark of quite how hot it is out there that Richard – who is usually pretty good in tropical climes – collapsed with heat stroke this afternoon. When he 'phoned, he sounded dreadful, and we would like to stress that the satellite 'phone link-up is of spectacularly poor quality, so although we guarantee that the gist of what is in this and future reports is correct, we are not certain that names of people or places are necessarily accurate.

Yesterday, we told how the team visited the home village of their guide and mentor Damon Corrie. We told how the 'phone cut out for a few seconds so we could not tell you the name of this village, but we asked Richard today and as far as we are able to ascertain, Damon's home village is called Pakuri (or something like that anyway).

Because communications are so difficult, we tried to have two or three people on the line listening every time a call comes through. This has been particularly irksome at times this evening when three of us answered the telephone at once, agog with anticipation, only to find that it was my beloved younger step-daughter wanting help with renewing her mobile 'phone contract. This is just one of the mildly amusing aspects of running the world's largest mystery animal research group from a family home.

Jon and Graham listened intently as the weak and garbled voice struggled through the ether. Hisses and pops punctuated the gems of information, which were, however, well worth the wait.

The team has spent today trekking deep into the grassland savannah. There is very little cover, and Jon Hare told us that the only thing that made the intense heat bearable were the occasional breezes. Jon told us that earlier on today they had found a small creek, and the whole team had waded in to cool down. He was particularly struck by the shoals of small fishes, which appeared – as if from nowhere – and immediately set to nibbling at the skin of our intrepid explorers. By the creek they met an old hunter. "Old" is a relative term. He was 53; no great age here in Britain – after all Corinna, Graham and I are rapidly approaching that venerable status, and Chris Clark on the expedition is quite a few years older than that. However, in a poorly developed third world country, 53 is quite a venerable age. Jon got talking to the hunter, who had tribal tattoos on his back, and told him how when he was young, at this very creek, and armed with only a machete, he had killed a jaguar.

They reached a village, which we think was called either Tolshiba (Graham's interpretation) or Calshida (Jon Downes'). Unfortunately, at this point, the line broke up. Richard told us how an eye-witness reported seeing the didi a couple of years ago. He described it as looking like an enormous white man covered in hair. At this point we cannot confirm whether this witness was at the village of ?Tolshiba/?Calshida, or whether this was an incident which took place earlier.

What we can confirm is that yesterday, at Letham Station, they met a number of eyewitnesses to giant anacondas. However, the largest of these was reportedly only 25 foot long. This is one hell of a snake, but it is just about within the accepted size range for this species. However, other eyewitnesses told them of tracks found in the jungle near water, which appear to be from much larger snakes. They intend to return to Letham Station at some point during the next fortnight and to visit the places where these snakes have been seen, with the eyewitnesses.

Jon Hare told us something very peculiar. Apparently, in the middle of the savannah, miles from any water there is a large flat rock known as 'Mermaid Rock'. Allegedly, within recent living memory, human figures with fishy tails have been seen sitting atop this rock performing the ultimate mermaid cliché of combing their hair. What mermaids would be doing so far from any water, beggars belief, but it is interesting that such a piece of archetypal European

folklore has been transplanted to a rock deep in the heart of nowhere.

Tomorrow they will continue travelling across the savannah. They intend to avoid the worst of the mid-day sun by travelling very early in the morning and in the later afternoon. Either tomorrow or the next day, they will reach a rocky area containing some ancient caves. These caves were allegedly inhabited by people eons ago. Jon Hare told us how they have been told stories of ancient human remains having been discovered deep in these caves. These bodies had been ritualistically interred in a similar way to some of the archaeological sites Jon has studied in Indonesia.

This is particularly interesting to cryptozoologists because, as Damon Corrie told us on the telephone last week, these caves have been reportedly the haunt of didi in recent years.

We await, with anticipation, the next report from the team."

Then there was another 24 hours of silence, and the burgeoning paranoia of the team back at Myrtle Cottage began to - once again - get out of hand.

We wrote:

"We have spent the whole day by the telephone waiting for news. But to no avail. We have regularly tried to telephone the guys out in Guyana, but each time we have not been able to get connected.

We will be manning the phones throughout the night, and will post a news bulletin as soon as we have any news to post. In the meantime, keep checking this page, and we will update you all as soon as we are able...."

It was another excruciating 24 hours before we heard anything else. By this time, the team at the CFZ HQ had evolved a good *modus operandi* with which to deal with the incoming calls from the SatPhone (which were becoming increasingly fractured and hard to interpret). All four of us (Graham, Corinna, Oll and me) would listen in on various extensions taking notes (in Corinna's case in shorthand), so that when we wrote up our account for the blog we could get some sort of consensus agreement on what it was Richard had actually said!

The first bulletin produced by the four of us was broadcast late at night on Monday 19th November, and as well as reporting the news of the previous few days, addressed a few ill-informed comments upon our previous postings which had appeared on one of the world's foremost cryptozoological websites:

"After nearly 48 hours of silence, Richard telephoned us at about 5.45 GMT, and – in order to conserve their dwindling resources of credit – we telephoned them back. They are having a particularly rough time with the heat; and Richard has once again suffered from heatstroke today. Lisa fell off a mountain and broke her thumb yesterday, and the other expedition members are also finding the terrain particularly gruelling.

"The heat is dreadful, I've never known heat like this", says Richard, who went on to say that they had seriously under-estimated the heat, the dryness, and the lack of shade. "It's relentlessly dry and hot with hardly any shade", he said, and his voice sounded fatigued and drawn.

However, they have had a very interesting couple of days. Yesterday they visited a cave system in some remote mountains. The connection of the telephone was particularly bad today and so, although he repeated it three times, we were unable to ascertain the name of the location, but, so it transpires, these cave systems were only discovered for the first time a few years ago. They contain a very ancient – and probably pre-Columbian – burial ground and they found graves of adults and children, including one where the child and adult had been buried in such a way that the adult appears to have been some kind of tribal shaman.

For those of you not aware of the term, the pre-Columbian era encompasses all periods of history and pre-history of the New World prior to the arrival of Columbus in 1492.

We are not in a position to give any more information about these burial grounds at this time. However, Richard tells us that the CFZ expedition were the first Europeans ever to visit these caves, and they have taken a lot of photographs and video film, which we will be broadcasting on CFZ upon their return. Expedition archaeologist Lisa Dowley – despite her broken thumb – took some samples, which will be examined by a British University upon their return to the UK. Once again, the quality of transmission was so poor that we cannot give any more details about what these samples are.

History does not relate whether Lisa had her accident before or after visiting these caves, and it has to be said – with tongue somewhat in cheek – that the Tintin adventure *Seven Crysal Balls/Prisoners of the Sun*, written by Georges Remi (aka Hergé) in the 1940s, comes to mind.

The main reason they visited these cave systems was that it has historically been known as a haunt of the didi. It was last seen ten years ago by a man who was so frightened that he ran away down the mountain. He described it as being a huge, hairy man. In the previous blog post we reported how another didi witness had described the creature he had seen as being like a European but covered in hair. When this was posted on one of the other internet blogs dealing with cryptozoology, someone made a comment that it could not possibly have been a European.

Well duh!

I think the important thing here is that, on the whole at least, people from Western Europe are far taller than people from South America. The description of the did as being 'like a European' is – in my opinion at least – merely a cute way of saying that the figure was tall and burly.

They also received some other strange stories of apparently hominoid creatures. They have received reports of a very small human-shaped animal; only two and a half feet tall, and with a bright red face, that was seen near these caves. The local people who told them of this believe that the red colouration had been painted on. They also heard of similar creatures seen nearby at a place called Trebang's Rock. Trebang are allegedly humanoid creatures, short in stature, which are – according to Richard – "supposed to touch children and transmit deadly diseases to them". He went on to say that when the children die, the Trebang is supposed to take their bones and make them into flutes.

South America has a long and fascinating history of small, often malevolent, dwarves. I would refer you to a book called *The Humanoids* by Charles Bowen (Spearman 1969) which, despite an appalling cover – which would lead many researchers to think that this is a tedious piece of sensationalist drivel – contains a wealth of fascinating, and unique information. I have been studying the mysterious dwarves of Central and South America for a long time, and although I believe that the paranormal attributes, which have often been given them by local folk, should be taken *cum grano salis*, I think that there may well be some serious cryptozoological evidence buried beneath a miasma of superstitious nonsense. Whether these 'little people' are human or animal, remains to be seen, although accounts like the one that the expedition team received today, suggest that they may well be very primitive human beings.

The team have travelled six miles today, so far under excruciatingly bad conditions. They are on their way towards some swamps where they will film anacondas and caymans. This is one of the few remaining strongholds for the increasingly endangered black cayman *(Melanosuchus niger)* and it will be a privilege for them to see this wonderful species in the wild. However, whilst in these swamps they will also be photographing and filming green anaconda *(Eunectes murinus)* in order to study the population pressure and feeding strategies of this species in this part of Guyana. The giant specimens have been reported at a place called Corona Falls, which is some seventy miles away from their present location. After a few days in these swamps, the team will be returning to Letham Station, and they hope to negotiate the hire of a helicopter to take them to the Falls. They hope that this will be within what remains of the expedition budget, although the paperwork is likely to be complicated because the company who owns the helicopter is based in Brazil.

All in all, an exciting, and eminently satisfactory two days for the team that Nick Redfern has dubbed 'the Guyana Five'. More news when we get it."

The next day we had another telephone call, but there was also more criticism from the inhabitants of cyberspace. The CFZ was only interested in getting a free holiday, they accused us, and others accused us of bringing the subject into disrepute by accepting sponsorship monies from a computer games manufacturer, and so the next news bulletin was not just a distillation of the news from the latest telephone call, but also a firm rebuttal of the nonsensical accusations:

"We heard from the increasingly beleaguered expedition, or, as Nick Redfern has dubbed them 'the Guyana five' just before 10 this evening GMT. Richard gave such a graphic description of the problems with which they have had to deal, that when he culminated his account with the quote that we have taken for the headline of this bulletin, we couldn't bring ourselves to reprimand him for having split an infinitive. After all – although I can feel my late father rolling in his grave as Corinna types this, there are worse things than massacring the Queen's English.

They reached the swamps late yesterday, and they have spent most of today exploring them.

In Richard's words they have been 'going through hell'. "No shade, no trees – it's a nightmare … the Gobi desert was a breeze compared to this", said Richard, and he described how the intense heat was causing a series of spontaneous bush fires. One of the biggest for the expedition is that, whereas on other expeditions they have started work at 7.00 in the morning and carried on to dusk, because of extreme conditions they just cannot do this. Richard described how he spent a large part of the hottest part of the day standing knee deep in a muddy swamp, in the inadequate shade of a crumbling tree, in a vain attempt to keep cool.

Perhaps 'armchair cryptozoologists', who like nothing more than to sit back and criticise expeditions such as this, would like to try really roughing it for once in their pampered lives. We have received a lot of mail over the past few days, most of it favourable and supportive, but there have been people whingeing about how the team are merely on an expensive foreign 'jolly'. Well, it was certainly expensive, they are certainly in foreign parts, but the descriptions of injuries, heat stroke, and other privations paint a far from jolly picture.

There have also been postings on one of the British Fortean message boards, suggesting that this expedition is fundamentally floored because it was paid for by a computer games manufacturer. This is, as Oll Lewis pointed out in a robust rebuttal, complete nonsense. These days nearly all scientific endeavours are sponsored by somebody; usually biochemical manufacturing companies, medical research facilities, but often more dubious portions of the military industrial complex. From where I am sitting, being sponsored by someone who is – as Berry Gordy once said – "part of the industry of human happiness", is morally, at least, infinitely preferable. People whom I regard as being morally dubious have offered us sponsorship in the past; people who wanted to use our research to bolster up their own peculiar political/religious agenda. We refused to have anything to do with them, and will continue to refuse to do so. We are proud to be carrying out this difficult, and without blowing our own trumpets too much, intrepid investigation in conjunction with Capcom, and hope that our relationship with this company, who do nothing worse than provide entertainment for millions of customers, will continue in the future.

Both Richard and Paul (Mr. Biffo) have suffered from heat stroke today, and on top of having broken her thumb two days ago, Lisa has now compounded her injuries by losing a toe nail, which has – in Richard's delightful West Midlands patois – "made her foot go all manky".

They spent the day battling with extreme heat in a vain search for anacondas. Although they didn't see any living snakes, they did find the tracks of a specimen that they estimated to be about 15 foot in length. This is one hell of a snake. For the record, the largest anaconda to be generally accepted by mainstream scientists was about 28 feet long, and the ones that the expedition is searching for evidence of, maybe between 40 and 50 feet in length, so the 15 footer that left tracks at Crane Pond, for our intrepid expedition to find is now't but a wee bairn. After fruitless investigation of Crane Pond, earlier today they trekked across the savannah to another pond called Cashew Pond. On their way they saw and filmed a giant anteater (Myrmecophaga tirdactyla) – a gloriously bizarre edentate which can grow to a size of 8 feet (2.4 metres).

This expedition is particularly exciting for me because my great hero Gerald Durrell visited Guyana when it was still known as British Guiana in 1950, and recounted his adventures in a wonderful book called *Three Singles to Adventure*. I suppose, here, I should reassure any of the readers of this blog who are agog with anticipation of some degree of sexual impropriety, that the term 'singles' refers to one-way rail tickets rather than any indication of their marital status. Coincidentally, on another trip three years later, Durrell too encountered a giant anteater, and brought her – hand tame – back to Britain where she captivated the hearts of both the book buying public and visitors to Paignton Zoo for many years. As I am feeling generous, to celebrate the end of the first half of the expedition, I will give a prize of a year's free membership of the CFZ to the first person who emails me on jon@cfz.org.uk telling me the anteater's name.

But I digress.

The team are staying at Cashew Pond tonight. As you read this, Richard will be going through agonies of guilt, because – as the expedition is living off the land – their hunter earlier shot a young cayman and they are planning to have it for their tea. As Richard is a great lover of the crocodile family, this will be extremely difficult for him. As I have eaten both alligator and crocodile in my time, I can reassure you – and by proxy them – that it tastes like a slightly fish version of McDonald's chicken nuggets, and will certainly not launch any of the team on to a full-time career as reptile gourmands.

Tomorrow they head back towards Letham Station, where, hopefully, they will be successful in chartering a helicopter to take them to Corona Falls, which is some 70 miles away, and is reputably the mother load as far as giant anacondas are concerned.

As Graham so rightly pointed out, this expedition really is experiencing "science in the raw", and I am sure, like us, your thoughts and prayers are with them."

For the record, I made somewhat of a cock-up during this blog posting. The answer that I was hoping for was 'Sarah Huggersack' - the delightful girl anteater who had captivated millions who read about her in Durrell's books, and later went to see her at Paignton Zoo. However, I should have read my Durrell more diligently, because it turned out that she was actually the second animal of her species caught by Durrell; the first had been a male called 'Amos'. So I ended up giving three prizes rather than one.

Then things got uncomfortable.

Steve Jones is an old friend of ours. He is a practicing pagan, but he is also a Justice of the Peace, and he is often, therefore, hailed as Britain's first Pagan Magistrate. He learned about a political problem in Guyana and wrote to us, concerned that the crisis might cause problems for the expedition.

Steve wrote:

"I noticed reading the latest news about Guyana on Google News Jon that things are kicking off with Venezuela and themselves over gold mining with an alleged border incursion by Venezuelan troops to blow up some dredges.I hope the CFZ are far enough away to avoid any unpleasantness kicking off.it also looks like their Chief justice and government are having problems.how much longer are the CFZ staying in the country?"

We did some investigation, including telephoning the Foreign Office, and decided that it was actually a fairly minor incident, and that there was nothing much to worry about. We posted the following:

"We have looked into the situation, and it appears that our team is safe. These alleged incursions are in a different part of the country, and also seem to be a semi-regular occurrence. There has been a boundary dispute with Venezuela for decades, but despite periodic upheavals like this latest one, the threat of all out war seems as distant as the possibility that the situation will reach an amicable conclusion.

But thanks Steve for letting us know..."

We also posted up the original news report for clarification:

Guyana, Venezuela at odds over gold boats

CARACAS, Venezuela, Nov. 17 (UPI) -- The destruction of two

gold-mining barges in the South American nation of Guyana has intensified a border dispute with neighboring Venezuela.

Guyana officials allege 36 Venezuelan soldiers were behind the destruction of the two gold-mining dredges in a disputed border region Thursday. The Guyana officials allege the Venezuelan soldiers used plastic explosives to blow up the equipment, the BBC reported Saturday.

Venezuelan Ambassador Dario Morandy said Friday his country's military had not violated Guyana's borders, adding the area where the dredges had been operating was owned by his country. "Venezuela was protecting its natural resources and we need to remove all illegal miners from the area," Morandy told Stabroek News. Guyana officials have opposed that stance, claiming the incident took place in the Cuyuni River, a region their country controls.

A formal protest of the incident has been filed by the Guyana Ministry of Foreign Affairs, and a group of military and police officials are expected to conduct an investigation to determine any wrongdoing.

Copyright 2007 by United Press International. All Rights Reserved.

However, all hell was let loose. Various fortean commentators started getting in on the act, and claimed that we were making light of the situation, and that we should withdraw the expedition forthwith. There was, in fact, no way that we could have done that, but still the voices of unreason railed at us across cyberspace.

A day or so later, Steve sent us this news report which suggested that the international political crisis was becoming defused. We posted it on the internet and hoped that, as a result, the people who were beginning to accuse me of putting the expedition members lives at risk, because of my megalomaniac desire to further the aims of the CFZ would just shut up.

The news report read:

Guyana, Venezuela agree to temporary "truce"

Following a verbal claim from Guyana Foreign Minister Rudolph Insanally in connection with a confusing border incident, Venezuelan Ambassador to Guyana Darío Morando said both countries agreed to wait until next November 22 to create a taskforce to assess the events related to the alleged blowout of two dredgers by the Venezuelan Army.

"We have met and are willing to find a diplomatic solution to this incident. Anyway, we have agreed to a truce until next Thursday, when Foreign Affairs Minister Nicolás Maduro is returning from an international trip," Morando explained.

He added that Maduro -who is accompanying President Hugo Chávez in a tour of the Middle East and Europe- need to be present in Venezuela to give the relevant instructions and designate the Venezuelan committee that is to deal with this issue. "The Ministers of Foreign Affairs could even meet, but I do not think this is necessary,"

Morandy added.

Guyana, however, has suggested the moves it is making if Venezuela does not provide a satisfactory answer. Georgetown plans to file an action with United Nations, Insanally said on Tuesday.

Last Thursday, two dredgers engaged in gold mining activities were blown out on the Venezuela-Guyana border. The Guyana government claims the incident came amidst a Venezuelan military operation allegedly intended to fight illegal mining operations and reportedly conducted in Guyana territory.

I decided, whilst I was certain that there was nothing to be worried about as far as the safety of the expedition from becoming collateral damage in what my detractors in cyberspace were already calling an `invasion` or even a `war`, although it was in fact no such thing, that I could not be seen to be taking the matter lightly. I therefore refrained from pointing out the delightful lexilink inherent in a politician being called Rudolph Insanally , and just sat back and hoped that the crisis would blow over.

To all intents and purposes it did.

Ironically, when the team got back to England, Richard laughingly accused me of trying to drum up publicity for the expedition by exaggerating a minor border incursion into a full-scale war in order to persuade newspapers to cover the expedition. Nothing could have been further from the truth, as you now know. Sometimes it seems that the more successful the CFZ becomes, the more people seem to be lining up to take a pop at us, and at me personally every time that I dare to put my head above the parapet.

There was another 48 hours of silence, and - considering what Richard had told me about the depredations of heatstroke amongst the tem, and Lisa's injuries - I was beginning to get worried again.

I am essentially a fairly paranoid and pessimistic person, but, being out of touch with

 CFZ GUYANA EXPEDITION 2007

the expeditionary team for so long would have worried anyone - not just a middle-aged cryptozoologist who has a tendency to become as mad as a bagful of cheese!

When the news finally came it was good, but I decided that I should include some reference to the political situation in my write-up, otherwise I would - once again - be setting myself up to be accused of being an uncaring megalomaniac:

"Finally! After a silence of 48 hours, which – considering the burgeoning political crisis in Guyana – I have to admit was being to make me seriously worried. So worried that – without meaning to – this must have communicated itself in my blog posts. We had carefully done our best not to put our worries into writing because some of the team have families back in the UK and two of them have small children. I am sure that said families are reading this blog for news of their loved ones and we did not want to cause undue alarm at such an early stage. Mrs. Biffo would never have forgiven us.

However, some of our concern must have leaked through, because at teatime this evening, we had a 'phone call from veteran cryptozoological explorer Adam Davies, who 'phoned to give us his support and offer any help that he could. God bless you Adam. We were fairly sure, however, that the information black-out was called by technical reasons, and the fact that this evening's bulletin was received in nearly a dozen tiny 'phone calls lasting less than a minute each, as the sat. 'phone kept on cutting out, would tend to corroborate our hypothesis.

But it was well worth the wait. Richard's first words to me we "Bloody hell, Jon, this place is a cryptozoological treasure trove!"

Sadly, however, because of the intermittent nature of the communications, although Richard had an amazing amount to tell us, it still left us with more questions than answers; so forgive us that there are some glaring gaps in this narrative, and that some of the questions that I know you will all want to be answered, remain, for the moment, a mystery.

The team limped back to Letham Station yesterday. They are all relatively all unscathed, although the telephone kept on cutting out as we asked for details of – for example – Lisa's condition. Back at Letham, they discovered some remarkable new information.

But the bad news first: they were unable to get hold of a helicopter at such short notice, and they are unable to reach the falls by boat, because it is the dry season and the river is too low. It is also too far too walk, and so, frustratingly, the trip to Corona Falls, where the enormous anacondas have been reported very recently will have to be postponed for a further expedition.

Richard told me: "It's absolutely bloody gutting to know that we are so near, but so far. However, the stuff we found out today more than makes up for it."

They are already planning a second trip to Guyana, this time in the rainy season, when the rivers will be navigable, it will not be so hot, and – from some of the anecdotal evidence they have garnered in the last 48 hours – it looks like they will have better luck in chasing some of the creatures that are their quarry.

For example: The water tiger. This is a particularly poorly known cryptid, and many researchers, including us, have suggested that a good identity for the beast would be a misidentification of the rare, and highly peculiar, South American giant otter *(Pteronura brasiliensis)*. However, it seems, from several pieces of anecdotal evidence that they have gathered in the last 48 hours, that this is quite simply not the case.

Yesterday they visited a township called Point Ranch. Kenard, one of their guides, introduced the team to a local man called Elmo. Elmo is very familiar with the water tiger and the giant otter. He says that the water tiger is a spotted animal with markings similar to those of a jaguar. They are aquatic, hunt in packs, and – somewhat peculiarly – he claims that the pack is led by an alpha animal that he refers to as 'the master', who orchestrates the hunting which is done by the younger members of the pack. Kenard apparently confirmed this statement, which – to the best of our knowledge – has never been published in Europe or the United States before.

Another local man told them that there is a mountain that is so remote that is doesn't have a name. No-one has ever climbed up it and come back alive. The team are not only too debilitated to climb it, but have decided that they would need special mountaineering equipment to do so. This is something that they intend to do on the next expedition, because according to their informant, there are water tigers up there, as well as a dragon-like creature who guards a spring. How these people know this, if no-one has ever come back alive seems somewhat of a moot point, but it would churlish to raise such objections when we have not heard anything like the full story. Richard assures us that all these interviews have been captured on video, and I am very much looking forward to the treat which awaits me, as I start editing the raw footage

into what will be the fourth of our major cryptozoological documentaries.
But there is more on the water tiger! Joseph one of their guides, told them how, back in the 1970s he had seen the pelt of a water tiger that had been shot. It was 10 feet long including the long tail, white – like a cow's hide – with black spots, and a striped head like that of a tiger. They have received reports of several different colour morphs of the water tiger, and Lisa has suggested that it could perhaps be some kind of huge aquatic mustelid, as not only are these animals known to – in some species at least – exhibit a wider range of colour varieties within a natural population than do most carnivores, but several species, including the European stoat *(Mustela erminea)* can change their colour according to the seasons. The stoat produces a white winter morph known as an ermine. Although as far as we are aware this has not been documented in a tropical species, it would seem perfectly feasible that, in a country where there is such a dichotomy between the rainy and dry seasons, that to have two seasonal colour morphs could well be a distinct evolutionary advantage.

But there is more!

Today they were back in Letham Station and they visited the Carapus Mountains. It has to be said at this point that the Google search reveals no such place and that, once again, there was dissention in the ranks between those of us who took notes on Richard's 'phone call – two of us thought it was `Caracus`, two of us thought it was `Carapus`, and Google hadn't heard of either. If anybody out there in Internet Land can help us with the spelling, we would be very grateful, although we are fairly convinced that we will have to wait until Richard and the gang return next week.

In the mountains they met a ex-chief who had retired some years ago, and now runs a fish farm. His name is Ernest, and ten years ago he saw an anaconda that he estimates as being 30 feet in length, in a pool about 30 miles from Letham. It was shot, and he claims that the skin was taken back to England. This – if it is true – would certainly have been done illegally, as it contravenes many international pieces of legislation. Somewhere there may well be a stupid bloody tourist with an inappropriate bloodlust who is sitting upon a piece of invaluable cryptozoological history. People like that make me almost livid with anger.

Ernest is now 59 years old; as we discussed the other day this is quite an age in an unforgiving country like Guyana. When he was 19 – in the late 1960s – he saw a tiny red-faced man very similar to the one we described in the posting of Monday 19th November. He described it as being 3ft tall, with a bright red face. The only difference between this account and the previous one which we published on Monday, is that Ernest believes that the red face is not war paint, but is part of its natural colouration, very similarly to one of the uakari monkeys of the genus Cacajao. The common name is believed to come from the indigenous name for 'Dutchman'; because the indigenous peoples found that the sight of badly sunburned Western Europeans was irresistibly reminiscent of these peculiar simians. The Fortean Omniverse is a particularly peculiar one, because the same patterns tend to occur and reoccur again and again. Across the world, wherever we have been on expeditions, even when we are looking for something completely different, we hear stories of dragons. And across the world we hear, again and again, peculiar local monkeys being compared unfavourably with immigrants from the Low Countries. The proboscis monkey of Indonesia is nicknamed orang or monyet belanda – meaning Dutch monkey or Dutchman – as the native Indonesians noticed that the Dutch colonisers also had a large belly and a protruding nose. Many apologies to any of our readers from the Netherlands, whom I am sure have neither bright red faces, large bellies, or ridiculous noses, but this was too choice an example of Fortean parasynchronicity to ignore.

These tiny "men" are said to like tobacco, and Ernest told the expedition members how, when the "little man" appeared, he gave it tobacco and it disappeared.

Damon chipped in, and said that something similar happened to him about 10 years ago. He was in a tent with his sister-in-law and another girl, when he awoke to see a red-faced man looking down at him. He was momentarily paralysed with fear, but then moved to try and protect the girls. When he turned round again, the figure had gone. He didn't hear the zip on the tent.

Ernest also told them that although he had not seen the didi (whereas we have always pronounced the word 'dee dee' it appears that it is actually pronounced 'die die') himself, he had heard of it. A friend of his, who only two years ago, saw a female didi in a tree nursing its child. He blacked out after watching it for some time, and when he had recovered it had gone. Soon after he became ill, and died within two years of seeing it.

This is a particularly interesting story because the inference is that this unnamed friend of Ernest died as a direct result of seeing this mysterious creature. We have heard of such things before. During the 2006 to The Gambia many of the people that the expedition interviewed said that to see a ninki nanka was fatal.

Kenard, their guide, told them that in the 1940s a local woman had been kidnapped by a didi, who took her away for several years, where she bore a child by him. So the story goes, when she finally managed to escape she swam across a river to a serendipitous hunter's boat, the hunter and the woman saw the didi brandishing its fist as if in grief on the bank of the river. It then picked up the hybrid child, and tore it apart.

This is another particularly widespread piece of folklore. We have heard similar stories about the yeti, the yeren, the almasty and possibly even bigfoot. Indeed, in the years before the gorilla was identified as a shy, gentle, and almost exclusively vegetarian great ape, similar stories of its sexual escapades returned to Europe in the form of traveller's tales.

Ernest told them of another potential cryptid – and this, to the best of our knowledge – has never been reported before in the annuls of cryptozoology. He is very familiar with Cuvier's dwarf cayman *(Paleosuchus palpebrosus)* – the smallest known species of the Alligatoridae, reaching a maximum size of a mere 1.5 metres. However, on two occasions, he has seen a tiny cayman, much smaller than the dwarf cayman, brown in colour, with a red stripe down its back. It bellowed loudly, and most peculiarly, he reported it has having two tails.

The expedition's driver said that he had seen these creatures as well, and Ernest took them to a cave system near a river where he claims that these creatures live.

The team explored these caves, and although they found nothing in there, Richard – who is, after all, a crocodilian expert, and was, at one time, Head of Reptiles at Twycross Zoo in the West Midlands of England - says that, in his opinion, these caves are eminently suitable for a small crocodilian to aestivate in during the harsh months of the dry summer. For those of you not in the known, aestivation is basically the polar opposite of hibernation; going into a semi-dormant or dormant state to escape extremes of hot weather.

We suggested that the seemingly insoluble problem of the creature being reported with two tails could perhaps be indicative of it not being a cayman at all, but being some kind of huge salamander. When the tails of salamanders and newts have been injured, they sometimes grow back double. But then again, so do those of some lizards, so for the moment this must remain an enigma. However, it is an enigma which we hope will not stay that way for long. Richard and the team are going back to see Ernest this evening for dinner and we hope that they will be able to get some more information from him.

Richard said that he had more to tell us, but at that point the telephone connection conked out for good. We tried to 'phone him back five or six times to no avail.

So, we were not able to find out the rest of Richard's exciting news. Nor were we able to reassure Mr. Biffo's many fans, some of whom have even eMailed us wanting to make sure that he was alright. We will try to do so tomorrow, but would ask everybody to be reassured that in a condition like this, no news is probably good news. If anything horrible had happened, Richard would, I am sure, have told us about it.

We also didn't have a chance to ask him what the local reaction was to the burgeoning political crisis in the country, if indeed a rural area so far from the capital was even aware of it.

Here ends a fascinating, though horribly frustrating blog entry. Hopefully, tomorrow, we will have some more news for you, and will be able to fill in some of the tantalising gaps that remain in today's narrative.

Thank you to all our readers – over a thousand a day – from all over the world, who are following the exploits of the Guyana five. Stick with us, it is going to be a bumpy ride.

And for those of you across the Atlantic from us in rural North Devon, a Happy Thanksgiving (CFZ HQ is only 50 miles from the original Plymouth Rock after all)."

Thanksgiving was something that we totally failed to take into consideration. The American holiday took place about half way through the expedition, and suddenly all interest in the expedition from the American media stopped, and viewing figures on the blog plummeted. I never realised quite how important a holiday it was in the USA, and that it was the season where everybody went to visit their Great Aunt Mabel, rather than sit at their computers and read news bulletins from a bunch of crazy limeys who were traversing a savannah in a South American country most of their peers had never heard of.

Then we had an eMail from Paul Rose. It appeared that he was back in Barbados and that the rest of the team were still in Guyana. We had no idea what had happened, so I spent a frantic 48 hours trying to contact him, before finally releasing the following note on the expedition blog:

"We are in email contact with Paul Rose. The team are on the way back. Paul has flown back to Barbados, while Richard, Lisa, Chris and Jon, together with their guides and hunters are apparently still in Guyana.

Details are scanty at the moment, but we will tell you more as we can. Paul writes:

"For me, the single weirdest thing has been the red-faced "bush people" who are pretty much accepted as fact by more or less everyone we met. While descriptions of the Dai Dai and the water tiger varied (though in the former case everyone described them as "big, hairy, men"), and the giant anaconda is a bit difficult to pin down (as it were), size-wise, the bush people were very consisent.

Even our guide, Damon, who hadn't really commented on anything monster-wise, eventually admitted to having seen one. And being a tribal chieftan he's very respectful of such things. Last night, while the others went to Pakuri, I stayed at the home/insect-ridden hovel of a guy called Marvin (neither paranoid, nor an android), who was an Arawak like Damon. He now lives in the centre of Georgetown, and has kind of turned his back on the whole Amerindian thing to a degree. I asked him about the Dai Dai, and just kind of laughed, and said that he doesn't think it exists (the first person we'd met to say that - but he was undoubtedly the most urbanised person I'd spoken to about it), but when I brought up the bush people he just kind of shrugged, and said "Yeah, they're real. They're like little, uncivilised people, who paint their faces red, like tobacco, and don't wear no clothes. Like pygmies".

Which I found rather interesting..."

He continues:

"It was absolutely the toughest thing I've ever done, certainly physically, and -

at times - emotionally too. Without question. The day we had on the savannah, following the anaconda/caiman hunt in the morning, was just unbearable, and the four mile-plus walk that evening - with heatstroke, nausea, dizziness, was the closest I've ever come to thinking I was going to die. Properly, genuinely, die. That and the fact that every vehicle we boarded seemed to break down in the middle of nowhere, in the blazing sun. Part of the reason we were all going to come back early is because the bus between Lethem and Georgetown is so desperately unreliable. They had to send out a replacement for us 90 minutes into our journey, and then the replacement nearly lost a wheel, and the drivers didn't have the correct tools to tighten it back on (hence we flew back from Lethem, so that we didn't miss our flights home - but the only flight we could get back was on Friday morning). It's a true third world country, with all the challenges that entails. Just the lack of proper roads, and the constant uneven surfaces, are killer. My virtually pristine, all-too-un-broken boots that I took out with me are coming back looking like they've been through a combine harvester.

The day we interviewed Ernesto Faris at his fish farm, and he took us to see the two-tailed red caiman cave was a particular highlight for me. Stepping into that jungle clearing, the caves in front of us, was pure The Lost World/Indiana Jones - vampire bats, anthills up to my shoulder, underground waterfalls. Incredible. And being the only people outside of Taushida to ever see the burial pots atop Taushida Mountain was pretty close to a highlight of my life. Could've done without being passed the skull of a tribal chieftain's son to hold, but it was a breathtaking moment. Nearly died on the way down, mind, but that was pretty much a daily occurrence for most of us.

Spirits remained mostly high, though we were all dreadfully fatigued towards the end. Not so much physically, just worn down overall I think. And the guys over there probably still are, I'm just lucky enough to have been able to get somewhere civilised (and have had enough spare money left to have afforded to do so), as a sort of decompression chamber before I re-enter the real world."

We were happier once we had spoken to him, because in the absence of any real news to the contrary, our collective imaginations had run riot, and we were beginning to imagine all sorts of horrific scenarios. A day later, Paul sent us this briefing, titled:

HOW I ENDED UP BACK IN BARBADOS

We had to weigh up the decision whether or not to get the bus (I use the term loosely - it's like something out of Mad Max) back from Lethem, to Georgetown. On the outward journey our first bus had broken down, and the replacement bus almost lost a wheel at one point. Plus, for ventilation the windows are kept open at all times, and when you're driving for 15 hours on a dust road you come out the other end looking like you've just been swept off to Oz in a tornado, sans the protective shell of a Kansas homestead. We didn't want to be flying for 12 hours in that condition - IF the bus had even got us to Georgetown on time. Which, Damon said, was doubtful.

We enquired about flights from Lethem's airfield, but the flight for the day we needed was booked up, and the only available flight was Friday. Just so happened that they had five seats (and space for a consignment of very noisy macaws). It was a tiny plane, and like every other vehicle we'd been in, it seemed to be falling apart. I was sat behind the pilot, and just as we took off, the entire back of his seat sheared off, and fell in my lap. I had to hold onto it for the duration of the journey.

Nobody wanted to go back - especially not as we'd just gotten into the rainforest, and started to hear tales of the Dai Dai in the mountains outside Lethem, and so we started talking about maybe getting our flights brought forward, and possibly having a day or two in Barbados, as a sort of reward for what we'd been through. Unfortunately, when we enquired about flights, there was only one seat on the plane out of Georgetown to Barbados (where our flight to Heathrow is booked to leave on Tuesday), and - once they'd done the sums - I was the only one left with enough money to pay the required fee.

I confess, the thought of seeing my kids again somewhat took the sting out of finishing the expedition before the others, but my insane and scary night on my own in Georgetown, and the fact I couldn't get my connecting flight to Heathrow moved forward, have put paid to that! Hey ho. Worse places to get stranded.

Damon very kindly arranged for the others to stay with his family in Pakuri (also known as St. Cuthbert's Mission), which was out of the ques-

red faced black spider monkey
Ateles paniscus

tion for me, as I'd given almost all of my camping gear to Damon as a thank you (and he was back in Lethem still - and I didn't want to stay in Georgetown if I could help it).

The guys started talking about whether they could follow up on the leads they'd got when we'd first been in Pakuri. We'd heard reports of a girl being abducted by a "big, hairy man", in a village 30 miles from Pakuri. Plus, they were planning to get some footage of anacondas, which had been seen in the grasslands around the village. This was their plan for the weekend, but we should know more on Tuesday, when they fly back. I'll drop you a line tomorrow with more.

And that was just about it.

When Richard got back to Georgetown he eMailed me a few more details:

"We are now in Georgetown.

Prior to flying back we met an old rubber tapper in the tiny airport at Lethem. He told us that he too had see one of the little red faced men or 'bush men'. He was a boy and was out hunting with his father (he is 59 now) when he saw a tiny red faced man looking at him through the undergrowth. He, unlike the other witnesses, thought that it had hairy skin. This may indicate it might have been wearing something rather than going naked as the other witnesses said. He also said it ran off on all fours. Could this have been a red faced black spider monkey? They don't occur in this part of the country now, as it is too arid. I suppose we will never know for sure.

Spent some time back in Damon's village. I spoke with Foster's father, Joseph. He had heard of the DiDi but had never seen one. He said that before the village was erected the red faced men were known in the area. They liked to wrestle humans and were very strong.

The way to defeat one was to topple it over as it had no knees and could not rise up again (once again the legs not bending motif). His father told him that a water tiger once lived in a cave in the area. It was large and dark furred. It was supposed to be dangerous. The largest

anaconda he had seen was 23 feet. It had been washed into the village by floods about ten years ago."
Then, two days later, it was all over.

Richard arrived back at CFZ HQ in North Devon late on the Wednesday night, and the next day, the business of sorting through all the data, photographs and film, began in earnest...

DOING IT FOR THE KIDS

After hearing of the trials and tribulations being endured by the Guyana expedition the regulars at the CFZ's local bar demonstrated their support for the intrepid team. The last Sunday of November saw an international flavour at the *Farmer's Arms*, where the punters wore Japanese tshirts promoting the expedition, and CAPCOM - the computer games company whose funding enabled 5 British lads and lasses (well, one lass) to explore the wildlands of Guyana.

Following news various injuries, including a broken thumb, heatstroke, and a "manky" foot, but celebrating the successes of the expedition, Allan and Jennie Lindsay - the landlord and landlady - pulled pints of lager for us back at base to drink to their health.

A number of locals enjoying a quiet Sunday afternoon drink were surprised when we arrived with a big box of tshirts but entered into the spirit of the thing and donned them.

However, the following day the kids of the village got in on the act.

As we told the world's press:

"In the North-Devonshire village of Woolsery there has only been one topic of conversation for the past few weeks; the Guyana expedition. Five intrepid monster hunters are currently enduring the baking heat and wild terrain of the South American country, rooting out witness reports of giant anacondas, the monstrous yeti-like dai-dai, a large water dwelling beast known as the water tiger, fearsome red-faced pygmies and a diminutive two-tailed caiman-like creature unheard of by anyone outside of Guyana before this expedition.

The expedition is being run by the Woolsery-based Centre for Fortean Zoology [CFZ] and when the young Braund-Philips brothers – Greg (9) Ross (12) and David (15) – read yesterday about how discussions regarding the expedition have taken over *The Farmer's Arms*, the village's pub, they wanted to get in on the action too.

Since the expedition set off, they have been avidly following the team's adventures on the expedition's blog and drawing pictures of the fascinating creatures they have read about. As well as the mysterious animals mentioned above, the boys were particularly excited to hear about how the explorers had taken the first video footage of a newly discovered green scorpion. As a result, Ross – a keen amateur naturalist - now wants to start keeping and breeding pet scorpions, although his parents might have other ideas!

The expedition was sponsored by video games manufacturer Capcom to tie-in with their new game Monster Hunter 2: Freedom for the Playstation Portable allowing the CFZ, the world's largest mystery animal research group, to undertake its most ambitious expedition to date.

L-r Ross, Dave and Greg Braund-Phillips

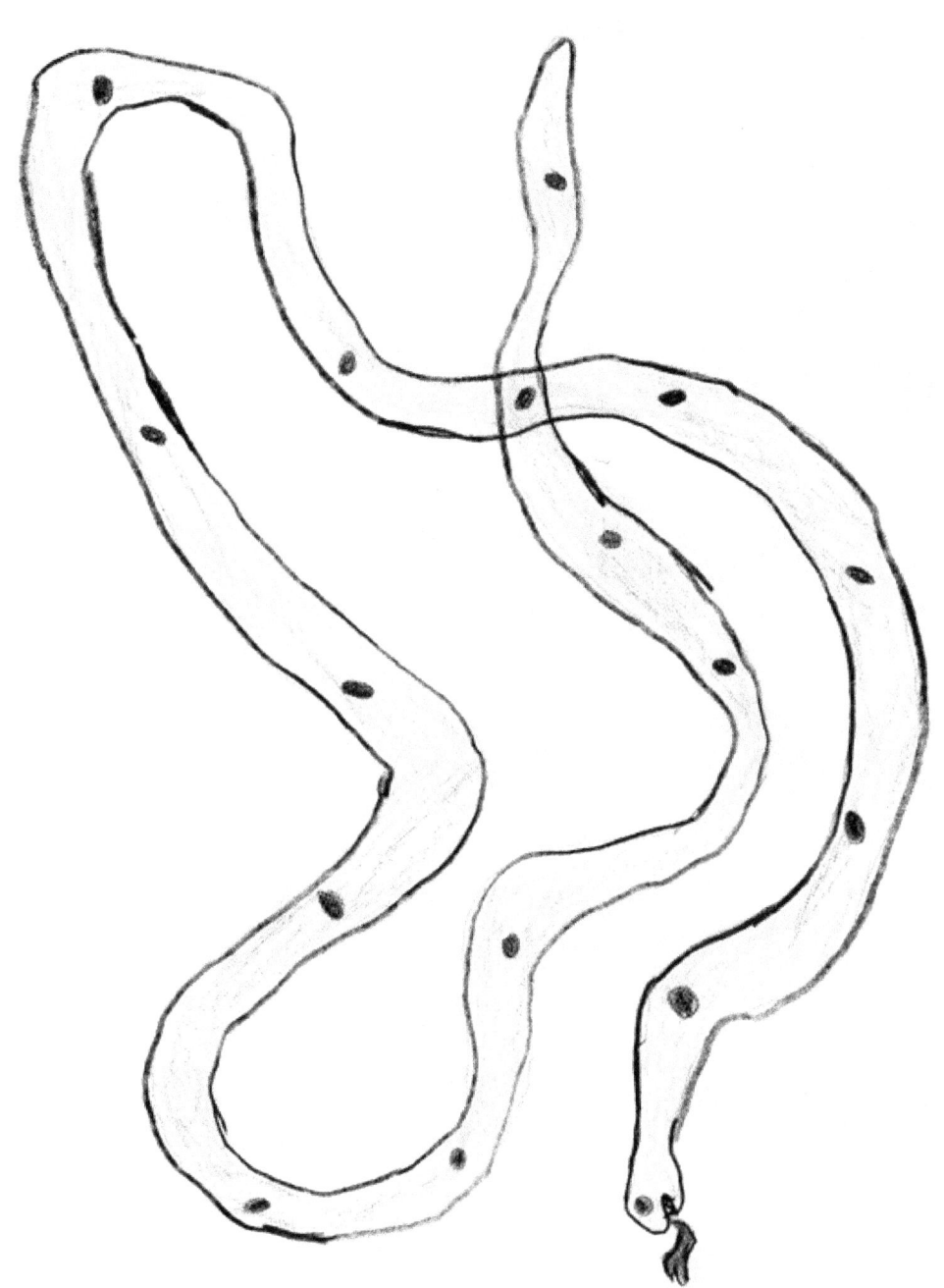

PRESS RELEASES

Myrtle Cottage, Woolfardisworthy, Bideford, North Devon EX39 5QR
Telephone 01237 431413
e-mail: cfz@eclipse.co.uk

For Immediate Release: 2007-10-24

On the 14th November 2007, five members of the Centre for Fortean Zoology – the world's largest organisation dedicated to the search for mystery animals – leave the UK for South America, on their most ambitious expedition yet. They will be searching the remote swamps and jungles of Guyana. They are looking for three elusive, potentially lethal, and hitherto undiscovered animals.

· The giant anaconda
· The didi
· The water tiger

As far as we are aware, this is the first cryptozoological expedition in search of evidence for the existence of these three animals that has ever been mounted. After months of complex negotiations, we can also announce that the expedition is sponsored by Capcom – one of the world's leading video game publishers, who are concurrently launching Monster Hunter Freedom 2, their exciting new game for the PlayStation 3 and PlayStation Portable (PSP).

The expedition will take the five members, and their guides, deep into unexplored swamps in the west of Guyana. The area is so remote and poorly known that it doesn't even have a name.

· The anaconda (Eunectes murinus) is the largest known snake in South America. The largest specimen shot was 28ft (9m) long. However, in the past, reports have come in from Guyana of anacondas of mind-boggling proportions, 40-60ft (12-18m) long. In some areas these giants are referred to as manatorro (the bull killer). As recently as last year, a specimen estimated at being 40ft (12m) long was observed by a party of native hunters. The giant snake frightened them so much that they fled. The target area for these monster serpents is a series of remote lakes in the grasslands.

· The didi is a more nebulous beast. It is said to walk upright like a man and be armed with scythe-like claws. It is alleged to tear out the tongues of living cattle, and leave swathes of terror in its wake. Although this last attribute may well be apocryphal, the claws in particular recall the supposedly extinct giant ground sloths or mylodonts. These bear-sized herbivores supposedly died out ten thousand years ago, but reports from across the Amazon, and surrounding areas, suggest they may well still survive.

· The water tiger is an aggressive aquatic animal said to have pointed teeth and webbed, humanlike hands. In the past, it was reported to have attacked both people and livestock. The water tiger may be based on reports of the rare giant otter (Pteronura brasiliensis) which can grow to a length of 6ft (1.8m).

The group intend to interview native witnesses to gather information on the animals and search the grasslands and lakes for evidence. They are being guided by Damon Corrie - a chief of the Eagle Clan Arawak tribe – who is also one of the few people to have visited the area in question.

The group consists of:

· Dr Chris Clark, engineer
· Lisa Dowley, photographer
· Richard Freeman, cryptozoologist
· Jon Hare, science writer
· Paul Rose, journalist

Photographs, a press pack, and further information are available, and expedition members are available for interview. Please contact Jon or Corinna at the CFZ Press Office on +44 (0)1237 431413.

NOTES FOR EDITORS:

+ The Centre for Fortean Zoology is a non profit-making organisation, which was founded in 1992. Over the last 11 years we have mounted expeditions to Central America, Thailand, Mexico, Mongolia, Sumatra, West Africa, various parts of the United States, as well as numerous investigations in the UK.
+ Further information on the CFZ can be found on their website, www.cfz.org.uk
+ CFZ Press are now the world's most prolific publishers of books on mystery animals.
+ The honorary life President of the Centre for Fortean Zoology is renowned explorer, author and soldier Colonel John Blashford-Snell OBE, best known for his pioneering Operation Drake and Operation Raleigh expeditions during the 1970s.
+ The CFZ is looking for corporate and private sponsors.
+ The CFZ make their own documentary films which can be seen at http://www.cfztv.org
+ `Lair of the Red Worm`, the 60 minute film of their 2005 expedition to Mongolia has now been seen by 27,000 people.

Myrtle Cottage, Woolfardisworthy, Bideford, North Devon EX39 5QR
Telephone 01237 431413
e-mail: cfz@eclipse.co.uk

FOR IMMEDIATE RELEASE – 12TH NOVEMBER 2007

FIVE EXPLORERS IN GIANT SNAKE ADVENTURE

On Wednesday evening, 14th November, a five-person expedition flies from Heathrow Airport in search of adventure. The five explorers from the UK based Centre for Fortean Zoology [CFZ] – the world's largest organisation dedicated to the search for unknown animals – are on the track of three potentially deadly monsters.

1. The giant anaconda. Although the largest known specimen of this snake was a mere 28 foot, there have been reports for centuries of far larger reptiles in the swamps of South America. Snakes measuring 40 – 50 feet have been reported from the trackless swamps of Guyana in the past few years. Are these creatures giant specimens of a known species? Or something entirely new? We aim to find out.

2. The didi. Often described as a bigfoot-type creature, some reports confuse it with the mapinguari; another huge South American mystery beast that some people believe is a surviving giant ground sloth. Giant ground sloth bear, hominid, or bogeyman? We aim to find out.

3. The water tiger. This poorly known aquatic beast is practically unknown in the west, but across South America it is famed for its ferocity. Is it something entirely new? Or is it something based upon sightings of the extremely rare giant river otter? We aim to find out.

The expedition will be keeping in touch with CFZ headquarters in rural North Devon by satellite 'phone, and CFZ Director Jonathan Downes (48) hopes to be able to post daily bulletins on a dedicated blog: http://cfzguyana.blogspot.com/

The expedition members are:
Richard Freeman, Zoological Director of the CFZ - expedition leader
Dr. Chris Clark, cryptozoologist
Lisa Dowley, photographer
Jon Hare, science writer and explorer
Paul Rose, journalist and author

Photographs of expedition members, a press pack, and other information are available. Please telephone Jon or Corinna on 01237 431413 for further details.

NOTES FOR EDITORS:

+ The Centre for Fortean Zoology is a non profit-making organisation, which was founded in 1992. Over the last 11 years we have mounted expeditions to Central America, Thailand, Mexico, Mongolia, Sumatra, West Africa, various parts of the United States, as well as numerous investigations in the UK.
+ Further information on the CFZ can be found on their website www.cfz.org.uk
+ CFZ Press are now the world's most prolific publishers of books on mystery animals.
+ The honorary life President of the Centre for Fortean Zoology is renowned explorer, author and soldier Colonel John Blashford-Snell OBE, best known for his pioneering Operation Drake and Operation Raleigh expeditions during the 1970s.
+ The CFZ is looking for corporate and private sponsors.
+ The CFZ make their own documentary films which can be seen at http://www.cfztv.org
+ `Lair of the Red Worm`, the 60 minute film of their 2005 expedition to Mongolia has now been seen by 27,000 people

Myrtle Cottage, Woolfardisworthy, Bideford, North Devon EX39 5QR
Telephone 01237 431413
e-mail: cfz@eclipse.co.uk

FOR IMMEDIATE RELEASE 23rd November 2007

GUYANA MONSTER HUNTERS 'PHONE HOME

Five British explorers from The Centre for Fortean Zoology [CFZ], based in rural North Devon, are deep in the little known grasslands of Guyana, South America on the track of unknown animals. The expedition is sponsored by Capcom, one of the world's leading video game manufacturers, and is led by Richard Freeman (37) a British zoologist and explorer.

The team have been in the wilds of South America for over a week now, and despite setbacks such as injury, heatstroke, equipment malfunction, and even a burgeoning political crisis which many commentators have described as an "invasion" of Guyana by neighbouring Venezuela, their initial findings appear remarkable.

They have obtained a wealth of anecdotal evidence for the existence of the three animals that they went there to investigate.

+ Didi
+ Giant anaconda
+ Water tiger

They have secured this evidence in the form of filmed interviews with eye-witnesses in this remote part of rural Guyana.

Amongst these are two chilling accounts of young women being abducted by the didi (pronounced 'die die'). None of these stories have been published in Europe or America before.

Perhaps the most important results so far are evidence for two completely hitherto unknown animals: a tiny brown crocodile, and a three-foot high hairy creature that walks upright like a man, and has a bright red face. They also have secured the first video footage of a living specimen of a recently discovered species of green scorpion.

The team do not return to the UK until Wednesday, 28th November, so there is plenty of time for you to join the thousands of people worldwide who follow their adventures on http://cfzguyana.blogspot.com/

CFZ Director Jonathan Downes is available for interview, and photographs of expedition members, and other images, are also available. Please telephone Jon or Corinna on 01237 431413.

NOTES FOR EDITORS

+ The Centre for Fortean Zoology [CFZ] is the world's largest mystery animal research organisation. It was founded in 1992 by British author Jonathan Downes (48) and is a non-profit making (not for profit) organisation registered with H.M. Stamp Office.
+ Life-president of the CFZ is Colonel John Blashford-Snell OBE, best known for his groundbreaking youth work organising the 'Operation Drake' and 'Operation Raleigh' expeditions in the 1970s and 1980s.
+ CFZ Director Jonathan Downes is the author and/or editor of over 20 books. Island of Paradise, his first hand account of two expeditions to the Caribbean island of Puerto Rico in search of the grotesque vampiric chupacabra, will be published in early 2008.
+ The CFZ have carried out expeditions across the world including Sumatra, Mongolia, Guyana, Gambia, Texas, Mexico, Thailand, Puerto Rico, Illinois, Loch Ness, and Loch Morar.
+ CFZ Press are the world's largest publishers of books on mystery animals. They also publish *Animals & Men*, the world's only cryptozoology magazine, and *Exotic Pets*, Britain's only dedicated magazine on the subject.
+ The CFZ produce their own full-length documentaries through their media division called CFZtv (http://www.cfztv.org/). One of their films *Lair of the Red Worm* which was released in early 2007 and documents their 2005 Mongolia expedition has now been seen by nearly 30,000 people.
+ The CFZ is based in Jon Downes' old family home in rural North Devon which he shares with his wife Corinna (51). It is also home to various members of the CFZ's permanent directorate and a collection of exotic animals.
+ Corinna and Jonathan Downes are shareholders in Tropiquaria – a small zoo in North Somerset (http://www.tropiquaria.co.uk/).
+ Jonathan Downes presents a monthly web TV show called On the Track (http://cfzmonthly.blogspot.com/) which covers cryptozoology and work of the CFZ.
+ The CFZ are currently building a Visitor Centre and Museum in Woolsery, North Devon.
+ Each year the CFZ presents an annual conference (http://www.weirdweekend.org/)
+ Following their successful partnership with Capcom (http://www.capcom.com/) on the 2007 Guyana expedition, the CFZ are looking for more commercial sponsors.

Myrtle Cottage, Woolfardisworthy, Bideford, North Devon EX39 5QR
Telephone 01237 431413
e-mail: cfz@eclipse.co.uk

FOR IMMEDIATE RELEASE: 2007-11-25

TINY RURAL VILLAGE SUPPORTS MONSTER HUNTER TEAM

After nearly a fortnight, the five person team from the Centre for Fortean Zoology [CFZ] – the world's largest mystery animal research group, are beginning their long journey home from the wilds of Guyana.

One member, Paul Rose, has already flown to Barbados, and the rest of the team are making their way back to the Guyanian cpital, Georgetown.

Initial reports from the team seem very exciting. They have obtained a wealth of anecdotal evidence for the existence of the three animals that they went there to investigate.+ Didi+ Giant anaconda+ Water tigerThey have secured this evidence in the form of filmed interviews with eye-witnesses in this remote part of rural Guyana.Amongst these are two chilling accounts of young women being abducted by the didi (pronounced 'die die'). None of these stories have been published in Europe or America before. Perhaps the most important results so far are evidence for two completely hitherto unknown animals: a tiny brown crocodile, and a three-foot high hairy creature that walks upright like a man, and has a bright red face. They also have secured the first video footage of a living specimen of a recently discovered species of green scorpion.

Over a thousand people a day have been checking into the expedition website http://cfzguyana.blogspot.com/ to keep up with news of the expedition, but the team also have vociferous support from people in the tiny North Devon village of Woolfardisworthy, where the CFZ is based.

Sunday lunchtime saw Allan and Jennie Lindsay, the landlord and landlady of The Farmers Arms, and their 8 year old daughter Clarissa, together with Angi Keen, the barmaid and quite a few of the customers, donning T Shirts commemorating the groundbreaking cryptozoological expedition, which is sponsored by Capcom – one of the world's leading video game manufacturers, to celebrate the release of a new game `Monster Hunter 2` for the PSP.

Images are available. Please telephone Jon or Corinna on 01237 431413.

NOTES FOR EDITORS

+ The Centre for Fortean Zoology [CFZ] is the world's largest mystery animal research organisation. It was founded in 1992 by British author Jonathan Downes (48) and is a non-profit making (not for profit) organisation registered with H.M. Stamp Office.
+ Life-president of the CFZ is Colonel John Blashford-Snell OBE, best known for his groundbreaking youth work organising the 'Operation Drake' and 'Operation Raleigh' expeditions in the 1970s and 1980s.
+ CFZ Director Jonathan Downes is the author and/or editor of over 20 books. *Island of Paradise*, his first hand account of two expeditions to the Caribbean island of Puerto Rico in search of the grotesque vampiric chupacabra, will be published in early 2008.
+ The CFZ have carried out expeditions across the world including Sumatra, Mongolia, Guyana, Gambia, Texas, Mexico, Thailand, Puerto Rico, Illinois, Loch Ness, and Loch Morar.+ CFZ Press are the world's largest publishers of books on mystery animals. They also publish *Animals & Men*, the world's only cryptozoology magazine, and *Exotic Pets*, Britain's only dedicated magazine on the subject.
+ The CFZ produce their own full-length documentaries through their media division called CFZtv (http://www.cfztv.org/). One of their films *Lair of the Red Worm* which was released in early 2007 and documents their 2005 Mongolia expedition has now been seen by nearly 30,000 people.
+ The CFZ is based in Jon Downes' old family home in rural North Devon which he shares with his wife Corinna (51). It is also home to various members of the CFZ's permanent directorate and a collection of exotic animals.
+ Corinna and Jonathan Downes are shareholders in Tropiquaria – a small zoo in North Somerset (http://www.tropiquaria.co.uk/).
+ Jonathan Downes presents a monthly web TV show called *On the Track* (http://cfzmonthly.blogspot.com/) which covers cryptozoology and work of the CFZ.+ The CFZ are currently building a Visitor Centre and Museum in Woolsery, North Devon.
+ Each year the CFZ presents an annual conference (http://www.weirdweekend.org/)
+ Following their successful partnership with Capcom (http://www.capcom.com/) on the 2007 Guyana expedition, the CFZ are looking for more commercial sponsors.

PRESS CUTTNGS

Monster hunters making hisstory

BY OLIVER STALLWOOD

THEY usually content themselves with hunting one terrifying beast per trip – but this time they are going after two.

Researchers from The Centre for Fortean Zoology (CFZ) are heading to Guyana in search of the didi – the South American version of Bigfoot – and giant anacondas.

In the past, the monster detectives have trekked through Mongolia to unearth the corrosive saliva-spitting 'Death Worm' and West Africa to trap the mythical ninki nanka.

So far, they have returned to their Devon base empty-handed, but this time the team say they are brimming with confidence.

Expedition organiser Jon Downes is organising the trip to a secret swamp in the rainforests of Guyana where there have been sightings of 12m (40ft) snakes.

He said: 'There have been reports of snakes that are much longer than any other.

'We believe these reptiles are either a new species altogether or extremely large anacondas.

'Whatever they are, it is it is going to be very exciting.'

The team will also go in search of

Slippery customer: A giant anaconda Picture: David Hamlin

the didi – a creature that lives 'somewhere between reality and folklore'.

It is described as a hairy, ape-like being which is believed to have killed hundreds of cattle by ripping out their tongues with its scythe-like claws.

The CFZ believes the beast may be a surviving species of ground sloth thought to have been extinct for more than 10,000 years. The expedition sets off in November and Metro reporter Oliver Stallwood will be in tow to document the team's progress.

It is being sponsored by Capcom, producers of Monster Hunter Freedom 2 – the fastest ever selling game on the PlayStation portable console.

Setting off in search of beasts

A UNIQUE group dedicated to the search of mystery animals is about to embark on its most ambitious expedition to date.

Members of the Centre for Fortean Zoology, based in Woolsery, will be leaving North Devon for South America next Wednesday.

They will be searching the remote swamps and jungles of Guyana looking for three elusive, potentially lethal, and undiscovered animals.

These include the Giant Anaconda, the Didi and the Water Tiger.

A spokesman said: "As far as we are aware, this is the first cryptozoological expedition in search of evidence for the existence of these three animals that has ever been mounted."

The Anaconda (Eunectes murinus) is said to be the largest known snake in South America. The largest specimen shot was 28ft (9m) long.

However, the team said in the past, reports have come in from Guyana of Anacondas of up to 40-60ft (12-18m) long.

Members said the Didi is a more "nebulous" beast and is said to walk upright like a man and be armed with scythe-like claws. It is alleged to tear out the tongues of living cattle, and leave swathes of terror in its wake.

The Water Tiger is described as an aggressive aquatic animal with pointed teeth and webbed, humanlike hands.

In the past, it was reported to have attacked both people and livestock.

The Water Tiger may be based on reports of the rare giant otter (Pteronura brasiliensis) which can grow to a length of 6ft (1.8m).

North Devon Journal
November 8th

MONSTER HUNTER

CE Europe Ltd
9th Floor
26-28 Hammersmith Grove   T +44 (0)20 8846 2550
Hammersmith, London W6 7HA   F +44 (0)20 8741 4176

Press Contact

Sam Brace (sam@capcomeuro.com)

Press information

Capcom join forces with The Centre for Fortean Zoology [CFZ] on a real life Monster Hunt to South America

London – 6th September 2007 – Capcom is thrilled to announce a new partnership with The Centre for Fortean Zoology [CFZ] the only professional scientific and full-time organisation in the world dedicated to cryptozoology - the study of unknown animals. Capcom is sponsoring the CFZ on a full scale Monster Hunt expedition to South America in association with its record breaking title *'Monster Hunter Freedom 2'*.

The expedition aims to locate and collect giant anaconda specimens with reports coming in of locals seeing the indigenous snakes as big as 40 feet long – twice as large as they're supposed to be – in nearby swampland. The CFZ hope to collect some young specimens and bring them back to their new zoo facility in Devon.

Monster Hunter Freedom 2 is a non-stop epic hunt-or-be-hunted 3rd person action adventure game and continues the Monster Hunter series with more content and options than ever before. Gamers increase their play skills as they battle through breath-taking environments, gathering rare flaura and fauna and seeking out mystical beasts whilst battling against bloodthirsty monsters in some of the best graphics ever seen on PSP system.

And just like in *Monster Hunter Freedom 2,* the expedition aims to look for evidence of a real life monster - the mythical 'didi', a feared South American devil described as a hairy ape-like being with scythe like claws and is believed responsible for the deaths of hundreds of cattle that were found with their tongues ripped out. The CFZ believe the beast may be a surviving species of ground sloth thought to have been extinct for over 10,000 years.

The expedition will take place over two weeks this November and will be made up of the CFZ specialist team of Monster Hunters. The team will be taken on a journey deep into the South American jungle by a shaman guide and will be posting daily blogs and footage of their adventures via their satellite enabled phones at http://cfzguyana.blogspot.com/. The last blog posted on a CFZ trip to Mongolia attracted over 70,000 unique impressions a day; the team hope to exceed that figure this time round.

The *Monster Hunter* franchise has become a cultural phenomenon in Japan with *Monster Hunter Freedom 2* notching up over 1.2 million sales since its release this April and holds the title of fastest ever selling game on the PlayStation portable console.

Other *Monster Hunter Freedom 2* Facts from Japan –

- Owners range from 8-50yrs old with an average age of 22.6 yrs old
- Owners of MHF2 spend on average 2-3 hours per day playing the game
- Owners have spent on average between a total of 200-300 hours playing the game since its release 6 months ago. However, some have clocked up an impressive 1000 hours
- An impressive 51.5% of MHF2 owners bought a PSP specifically to play MHF2
- Regional tournaments in 5 key cities attracted a total of 15,000 participants, but many had to be turned away due to oversubscription. Close to £100,000 of MH merchandise was sold at these events. Pictures are available on request.

Monster Hunter Freedom 2 releases in Europe on 7th September 2007 exclusively for Sony PlayStation Portable system.

Capcom is a leading worldwide developer, publisher and distributor of interactive entertainment. Founded in 1983, the company has created world renowned franchises including *Resident Evil, Street Fighter, Mega Man, Breath of Fire, Devil May Cry* and the *Onimusha* series. Headquartered in Osaka, Japan, the company maintains operations in the U.S., United Kingdom, Germany, Tokyo and Hong Kong. More information about Capcom and its products can be found on the company's web site at www.capcom.com.

###

Capcom, Mega Man, Resident Evil, Onimusha, Devil May Cry and Breath of Fire are either registered trademarks or trademarks of Capcom Co., Ltd., in the U.S. or other countries. Street Fighter is a registered trademark of Capcom U.S.A., Inc. "PlayStation" and the "PS" Family logo are registered trademarks of Sony Computer Entertainment Inc. Microsoft, Xbox and the Xbox Logos are either registered trademarks or trademarks of Microsoft Corporation in the U.S. and/or other countries and are used under license from Microsoft. All rights reserved. Tomb Raider and Lara Croft are registered trademarks of Core Design Ltd. All other marks are the property of their respective owners.

CE Europe Ltd

9th Floor
26-28 Hammersmith Grove   T +44 (0)20 8846 2550
Hammersmith, London W6 7HA   F +44 (0)20 8741 4176

Press Contact

Sam Brace (sam@capcomeuro.com)

Press information

capcomeuro-press.com

Founded in Japan in 1979 as a manufacturer and distributor of electronic game machines, Capcom (short for Capsule-Computer) has built a reputation for introducing cutting-edge technology and software to the video game market. A leader in the video game industry for 20 years, Capcom's legacy of historic franchises in home and arcade gaming are testament to an unparalleled commitment to excellence.

Building on its origins as a game machine manufacturer, Capcom is now involved in all areas of the video game industry and has offices in Tokyo and Osaka, Japan; California, USA; London, England; Hamburg, Germany, and Hong Kong, China. Capcom plan to open new offices throughout European territories as it continues to expand outside its traditional heartland of Japan.

Blockbuster franchises like **Resident Evil, Street Fighter, Devil May Cry** and the ever-popular **Mega Man** series set the standard in creative innovation, character development and unsurpassed gameplay. With unlimited creativity and technical expertise, Capcom continues to develop blockbuster hit after blockbuster hit and is synonymous with outstanding control, vibrant graphics and unrivalled playability.

Capcom develops products for all age groups incorporating games aimed at the family such as the **Mega Man** series, **Zack and Wiki** and the **MotoGP** series, to games aimed at the more adult market with titles such as **Dead Rising, Lost Planet, Resident Evil, Devil May Cry** and the **Onimusha** series.

www.capcom.com

Capcom, Mega Man, Resident Evil, Onimusha, Devil May Cry and Dead Rising are either registered trademarks or trademarks of Capcom Co., Ltd., in the U.S. or other countries. Street Fighter is a registered trademark of Capcom U.S.A., Inc. "PlayStation" and the "PS" Family logo are registered trademarks of Sony Computer Entertainment Inc. All rights reserved. All other marks are the property of their respective owners.

THE CENTRE FOR FORTEAN ZOOLOGY

So, what is the Centre for Fortean Zoology?

We are a non profit-making organisation founded in 1992 with the aim of being a clearing house for information, and coordinating research into mystery animals around the world. We also study out of place animals, rare and aberrant animal behaviour, and Zooform Phenomena; – little-understood "things" that appear to be animals, but which are in fact nothing of the sort, and not even alive (at least in the way we understand the term).

Why should I join the Centre for Fortean Zoology?

Not only are we the biggest organisation of our type in the world but - or so we like to think - we are the best. We are certainly the only truly global Cryptozoological research organisation, and we carry out our investigations using a strictly scientific set of guidelines. We are expanding all the time and looking to recruit new members to help us in our research into mysterious animals and strange creatures across the globe. Why should you join us? Because, if you are genuinely interested in trying to solve the last great mysteries of Mother Nature, there is nobody better than us with whom to do it.

What do I get if I join the Centre for Fortean Zoology?

For £12 a year, you get a four-issue subscription to our journal *Animals & Men*. Each issue contains 60 pages packed with news, articles, letters, research papers, field reports, and even a gossip column! The magazine is A5 in format with a full colour cover. You also have access to one of the world's largest collections of resource material dealing with cryptozoology and allied disciplines, and people from the CFZ membership regularly take part in fieldwork and expeditions around the world.

How is the Centre for Fortean Zoology organized?

The CFZ is managed by a three-man board of trustees, with a non-profit making trust registered with HM Government Stamp Office. The board of trustees is supported by a Permanent Directorate of full and part-time staff, and advised by a Consultancy Board of specialists - many of whom who are world-renowned experts in their particular field. We have regional representatives across the UK, the USA, and many other parts of the world, and are affiliated with other organisations whose aims and protocols mirror our own.

I am new to the subject, and although I am interested I have little practical knowledge. I don't want to feel out of my depth. What should I do?

Don't worry. We were *all* beginners once. You'll find that the people at the CFZ are friendly and approachable. We have a thriving forum on the website which is the hub of an ever-growing electronic community. You will soon find your feet. Many members of the CFZ Permanent Directorate started off as ordinary members, and now work full time chasing monsters around the world.

I have an idea for a project which isn't on your website. What do I do?

Write to us, e-mail us, or telephone us. The list of future projects on the website is not exhaustive. If you have a good idea for an investigation, please tell us. We may well be able to help.

How do I go on an expedition?

We are always looking for volunteers to join us. If you see a project that interests you, do not hesitate to get in touch with us. Under certain circumstances we can help provide funding for your trip. If you look on the future projects section of the website, you can see some of the projects that we have pencilled in for the next few years.

In 2003 and 2004 we sent three-man expeditions to Sumatra looking for Orang-Pendek - a semi-legendary bipedal ape. The same three went to Mongolia in 2005. All three members started off merely subscribers to the CFZ magazine.

Next time it could be you!

Project Kerinci, Sumatra - 2003
In search of the bipedal ape Orang Pendek

How is the Centre for Fortean Zoology funded?

We have no magic sources of income. All our funds come from donations, membership fees, works that we do for TV, radio or magazines, and sales of our publications and merchandise. We are always looking for corporate sponsorship, and other sources of revenue. If you have any ideas for fund-raising please let us know. However, unlike other cryptozoological organisations in the past, we do not live in an intellectual ivory tower. We are not afraid to get our hands dirty, and furthermore we are not one of those organisations where the membership have to raise money so that a privileged few can go on expensive foreign trips. Our research teams both in the UK and abroad, consist of a mixture of experienced and inexperienced personnel. We are truly a community, and work on the premise that the benefits of CFZ membership are open to all.

What do you do with the data you gather from your investigations and expeditions?

Reports of our investigations are published on our website as soon as they are available. Preliminary reports are posted within days of the project finishing.

Each year we publish a 200 page yearbook containing research papers and expedition reports too long to be printed in the journal. We freely circulate our information to anybody who asks for it.

Is the CFZ community purely an electronic one?

No. Each year since 2000 we have held our annual convention - the *Weird Weekend* - in Exeter. It is three days of lectures, workshops, and excursions. But most importantly it is a chance for members of the CFZ to meet each other, and to talk with the members of the permanent directorate in a relaxed and informal setting and preferably with a pint of beer in one hand. Starting this year - 18-20 August 2006 - the *Weird Weekend* will be bigger and better and held in the idyllic rural location of Woolsery in North Devon.

We are hoping to start up some regional groups in both the UK and the US which will have regular meetings, work together on research projects, and maybe have a mini convention of their own.

Since relocating to North Devon in 2005 we have become ever more closely involved with other community organisations, and we hope that this trend will continue. We also work closely with Police Forces across the UK as consultants for animal mutilation cases, and during 2006 we intend to forge closer links with the coastguard and other community services. We want to work closely with those who regularly travel into the Bristol Channel, so that if the recent trend of exotic animal visitors to our coastal waters continues, we can be out there as soon as possible.

We are building a Visitor's Centre in rural North Devon. This will not be open to the general public, but will provide a museum, a library and an educational resource for our members (currently over 400) across the globe. We are also planning a youth organisation which will involve children and young people in our activities.

Apart from having been the only Fortean Zoological organisation in the world to have consistently published material on all aspects of the subject for over a decade, we have achieved the following concrete results:

- Disproved the myth relating to the headless so-called sea-serpent carcass of Durgan beach in Cornwall 1975
- Disproved the story of the 1988 puma skull of Lustleigh Cleave
- Carried out the only in-depth research ever into mythos of the Cornish Owlma
- Made the first records of a tropical species of lamprey
- Made the first records of a luminous cave gnat larva in Thailand.
- Discovered a possible new species of British mammal - The Beech Marten.
- In 1994-6 carried out the first archival fortean zoological survey of Hong Kong.
- In the year 2000, CFZ theories where confirmed when an entirely new species of lizard was found resident in Britain.
- Identified the monster of Martin Mere in Lancashire as a giant wels catfish
- Expanded the known range of Armitage's skink in the Gambia by 80%
- Obtained photographic evidence of the remains of Europe's largest known pike
- Carried out the first ever in-depth study of the *ninki-nanka*
- Carried out the first attempt to breed Puerto Rican cave snails in captivity
- Were the first European explorers to visit the `lost valley` in Sumatra

EXPEDITIONS & INVESTIGATIOINS TO DATE INCLUDE

- 1998 Puerto Rico, Florida, Mexico *(Chupacabras)*
- 1999 Nevada *(Bigfoot)*
- 2000 Thailand *(Giant snakes called nagas)*
- 2002 Martin Mere *(Giant catfish)*
- 2002 Cleveland *(Wallaby mutilation)*
- 2003 Bolam Lake *(BHM Reports)*
- 2003 Sumatra *(Orang Pendek)*
- 2003 Texas *(Bigfoot; giant snapping turtles)*
- 2004 Sumatra *(Orang Pendek; cigau, a sabre-toothed cat)*
- 2004 Illinois *(Black panthers; cicada swarm)*
- 2004 Texas *(Mystery blue dog)*
- 2004 Puerto Rico *(Chupacabras; carnivorous cave snails)*
- 2005 Belize *(Affiliate expedition for hairy dwarfs)*
- 2005 Mongolia *(Allghoi Khorkhoi aka Mongolian death worm)*
- 2006 Gambia *(Gambo - Gambian sea monster , Ninki Nanka and Armit-   age s skink*
- 2006 Llangorse Lake *(Giant pike, giant eels)*
- 2006 Windermere *(Giant eels)*
- 2007 Coniston Water *(Giant eels)*
- 2007 Guyana *(Giant anaconda, didi, water tiger)*

To apply for a <u>FREE</u> information pack about the organisation and details of how to join, plus information on current and future projects, expeditions and events.

Send a stamped and addressed envelope to:

**THE CENTRE FOR FORTEAN ZOOLOGY
MYRTLE COTTAGE, WOOLSERY,
BIDEFORD, NORTH DEVON
EX39 5QR.**

or alternatively visit our website at:
www.cfz.org.uk

Other books available from
CFZ PRESS

CFZ PRESS

THE OWLMAN AND OTHERS - 30th Anniversary Edition
Jonathan Downes - ISBN 978-1-905723-02-7

£14.99

EASTER 1976 - Two young girls playing in the churchyard of Mawnan Old Church in southern Cornwall were frightened by what they described as a "nasty bird-man". A series of sightings that has continued to the present day. These grotesque and frightening episodes have fascinated researchers for three decades now, and one man has spent years collecting all the available evidence into a book. To mark the 30th anniversary of these sightings, Jonathan Downes has published a special edition of his book.

DRAGONS - More than a myth?
Richard Freeman - ISBN 0-9512872-9-X

£14.99

First scientific look at dragons since 1884. It looks at dragon legends worldwide, and examines modern sightings of dragon-like creatures, as well as some of the more esoteric theories surrounding dragonkind.

Dragons are discussed from a folkloric, historical and cryptozoological perspective, and Richard Freeman concludes that: "When your parents told you that dragons don't exist - they lied!"

MONSTER HUNTER
Jonathan Downes - ISBN 0-9512872-7-3

£14.99

Jonathan Downes' long-awaited autobiography, *Monster Hunter*...

Written with refreshing candour, it is the extraordinary story of an extraordinary life, in which the author crosses paths with wizards, rock stars, terrorists, and a bewildering array of mythical and not so mythical monsters, and still just about manages to emerge with his sanity intact.......

MONSTER OF THE MERE
Jonathan Downes - ISBN 0-9512872-2-2

£12.50

It all starts on Valentine's Day 2002 when a Lancashire newspaper announces that "Something" has been attacking swans at a nature reserve in Lancashire. Eyewitnesses have reported that a giant unknown creature has been dragging fully grown swans beneath the water at Martin Mere. An intrepid team from the Exeter based Centre for Fortean Zoology, led by the author, make two trips – each of a week – to the lake and its surrounding marshlands. During their investigations they uncover a thrilling and complex web of historical fact and fancy, quasi Fortean occurrences, strange animals and even human sacrifice.

**CFZ PRESS, MYRTLE COTTAGE,
WOOLFARDISWORTHY BIDEFORD,
NORTH DEVON, EX39 5QR
w w w . c f z . o r g . u k**

Other books available from
CFZ PRESS

ONLY FOOLS AND GOATSUCKERS
Jonathan Downes - ISBN 0-9512872-3-0

£12.50

In January and February 1998 Jonathan Downes and Graham Inglis of the Centre for Fortean Zoology spent three and a half weeks in Puerto Rico, Mexico and Florida, accompanied by a film crew from UK Channel 4 TV. Their aim was to make a documentary about the terrifying chupacabra - a vampiric creature that exists somewhere in the grey area between folklore and reality. This remarkable book tells the gripping, sometimes scary, and often hilariously funny story of how the boys from the CFZ did their best to subvert the medium of contemporary TV documentary making and actually do their job.

WHILE THE CAT'S AWAY
Chris Moiser - ISBN: 0-9512872-1-4

£7.99

Over the past thirty years or so there have been numerous sightings of large exotic cats, including black leopards, pumas and lynx, in the South West of England. Former Rhodesian soldier Sam McCall moved to North Devon and became a farmer and pub owner when Rhodesia became Zimbabwe in 1980. Over the years despite many of his pub regulars having seen the "Beast of Exmoor" Sam wasn't at all sure that it existed. Then a series of happenings made him change his mind. Chris Moiser—a zoologist—is well known for his research into the mystery cats of the westcountry. This is his first novel.

CFZ EXPEDITION REPORT 2006 - GAMBIA
ISBN 1905723032

£12.50

In July 2006, The J.T.Downes memorial Gambia Expedition - a six-person team - Chris Moiser, Richard Freeman, Chris Clarke, Oll Lewis, Lisa Dowley and Suzi Marsh went to the Gambia, West Africa. They went in search of a dragon-like creature, known to the natives as `Ninki Nanka`, which has terrorized the tiny African state for generations, and has reportedly killed people as recently as the 1990s. They also went to dig up part of a beach where an amateur naturalist claims to have buried the carcass of a mysterious fifteen foot sea monster named 'Gambo', and they sought to find the Armitage's Skink (*Chalcides armitagei*) - a tiny lizard first described in 1922 and only rediscovered in 1989. Here, for the first time, is their story.... With an forward by Dr. Karl Shuker and introduction by Jonathan Downes.

BIG CATS IN BRITAIN YEARBOOK 2006
Edited by Mark Fraser - ISBN 978-1905723-01-0

£10.00

Big cats are said to roam the British Isles and Ireland even now as you are sitting and reading this. People from all walks of life encounter these mysterious felines on a daily basis in every nook and cranny of these two countries. Most are jet-black, some are white, some are brown, in fact big cats of every description and colour are seen by some unsuspecting person while on his or her daily business. 'Big Cats in Britain' are the largest and most active group in the British Isles and Ireland This is their first book. It contains a run-down of every known big cat sighting in the UK during 2005, together with essays by various luminaries of the British big cat research community which place the phenomenon into scientific, cultural, and historical perspective.

**CFZ PRESS, MYRTLE COTTAGE,
WOOLSERY, BIDEFORD,
NORTH DEVON, EX39 5QR
w w w . c f z . o r g . u k**

Other books available from
CFZ PRESS

THE SMALLER MYSTERY CARNIVORES OF THE WESTCOUNTRY
Jonathan Downes - ISBN 978-1-905723-05-8

£7.99

Although much has been written in recent years about the mystery big cats which have been reported stalking Westcountry moorlands, little has been written on the subject of the smaller British mystery carnivores. This unique book redresses the balance and examines the current status in the Westcountry of three species thought to be extinct: the Wildcat, the Pine Marten and the Polecat, finding that the truth is far more exciting than the currently held scientific dogma. This book also uncovers evidence suggesting that even more exotic species of small mammal may lurk hitherto unsuspected in the countryside of Devon, Cornwall, Somerset and Dorset.

THE BLACKDOWN MYSTERY
Jonathan Downes - ISBN 978-1-905723-00-3

£7.99

Intrepid members of the CFZ are up to the challenge, and manage to entangle themselves thoroughly in the bizarre trappings of this case. This is the soft underbelly of ufology, rife with unsavoury characters, plenty of drugs and booze." That sums it up quite well, we think. A new edition of the classic 1999 book by legendary fortean author Jonathan Downes. In this remarkable book, Jon weaves a complex tale of conspiracy, anti-conspiracy, quasi-conspiracy and downright lies surrounding an air-crash and alleged UFO incident in Somerset during 1996. However the story is much stranger than that. This excellent and amusing book lifts the lid off much of contemporary forteana and explains far more than it initially promises.

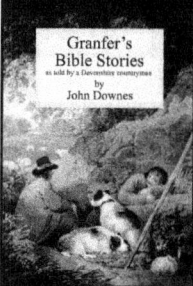

GRANFER'S BIBLE STORIES
John Downes - ISBN 0-9512872-8-1

£7.99

Bible stories in the Devonshire vernacular, each story being told by an old Devon Grandfather - 'Granfer'. These stories are now collected together in a remarkable book presenting selected parts of the Bible as one more-or-less continuous tale in short 'bite sized' stories intended for dipping into or even for bed-time reading. `Granfer` treats the biblical characters as if they were simple country folk living in the next village. Many of the stories are treated with a degree of bucolic humour and kindly irreverence, which not only gives the reader an opportunity to re-evaluate familiar tales in a new light, but do so in both an entertaining and a spiritually uplifting manner.

FRAGRANT HARBOURS DISTANT RIVERS
John Downes - ISBN 0-9512872-5-7

£12.50

Many excellent books have been written about Africa during the second half of the 19th Century, but this one is unique in that it presents the stories of a dozen different people, whose interlinked lives and achievements have as many nuances as any contemporary soap opera. It explains how the events in China and Hong Kong which surrounded the Opium Wars, intimately effected the events in Africa which take up the majority of this book. The author served in the Colonial Service in Nigeria and Hong Kong, during which he found himself following in the footsteps of one of the main characters in this book; Frederick Lugard – the architect of modern Nigeria.

**CFZ PRESS, MYRTLE COTTAGE,
WOOLFARDISWORTHY BIDEFORD,
NORTH DEVON, EX39 5QR
www.cfz.org.uk**

Other books available from
CFZ PRESS

ANIMALS & MEN - Issues 1 - 5 - In the Beginning
Edited by Jonathan Downes - ISBN 0-9512872-6-5

£12.50

At the beginning of the 21st Century monsters still roam the remote, and sometimes not so remote, corners of our planet. It is our job to search for them. The Centre for Fortean Zoology [CFZ] is the only professional, scientific and full-time organisation in the world dedicated to cryptozoology - the study of unknown animals. Since 1992 the CFZ has carried out an unparalleled programme of research and investigation all over the world. We have carried out expeditions to Sumatra (2003 and 2004), Mongolia (2005), Puerto Rico (1998 and 2004), Mexico (1998), Thailand (2000), Florida (1998), Nevada (1999 and 2003), Texas (2003 and 2004), and Illinois (2004). An introductory essay by Jonathan Downes, notes putting each issue into a historical perspective, and a history of the CFZ.

ANIMALS & MEN - Issues 6 - 10 - The Number of the Beast
Edited by Jonathan Downes - ISBN 978-1-905723-06-5

£12.50

At the beginning of the 21st Century monsters still roam the remote, and sometimes not so remote, corners of our planet. It is our job to search for them. The Centre for Fortean Zoology [CFZ] is the only professional, scientific and full-time organisation in the world dedicated to cryptozoology - the study of unknown animals. Since 1992 the CFZ has carried out an unparalleled programme of research and investigation all over the world. We have carried out expeditions to Sumatra (2003 and 2004), Mongolia (2005), Puerto Rico (1998 and 2004), Mexico (1998), Thailand (2000), Florida (1998), Nevada (1999 and 2003), Texas (2003 and 2004), and Illinois (2004). Preface by Mark North and an introductory essay by Jonathan Downes, notes putting each issue into a historical perspective, and a history of the CFZ.

BIG BIRD! Modern Sightings of Flying Monsters

Ken Gerhard - ISBN 978-1-905723-08-9

£7.99

From all over the dusty U.S./Mexican border come hair-raising stories of modern day encounters with winged monsters of immense size and terrifying appearance. Further field sightings of similar creatures are recorded from all around the globe. What lies behind these weird tales? Ken Gerhard is a native Texan, he lives in the homeland of the monster some call 'Big Bird'. Ken's scholarly work is the first of its kind. On the track of the monster, Ken uncovers cases of animal mutilations, attacks on humans and mounting evidence of a stunning zoological discovery ignored by mainstream science. Keep watching the skies!

STRENGTH THROUGH KOI
They saved Hitler's Koi and other stories

£7.99

Jonathan Downes - ISBN 978-1-905723-04-1

Strength through Koi is a book of short stories - some of them true, some of them less so - by noted cryptozoologist and raconteur Jonathan Downes. The stories are all about koi carp, and their interaction with bigfoot, UFOs, and Nazis. Even the late George Harrison makes an appearance. Very funny in parts, this book is highly recommended for anyone with even a passing interest in aquaculture, but should be taken definitely *cum grano salis*.

CFZ PRESS, MYRTLE COTTAGE, WOOLSERY, BIDEFORD, NORTH DEVON, EX39 5QR

Other books available from
CFZ PRESS

CFZ PRESS

BIG CATS IN BRITAIN YEARBOOK 2007
Edited by Mark Fraser - ISBN 978-1-905723-09-6

£12.50

People from all walks of life encounter mysterious felids on a daily basis, in every nook and cranny of the UK. Most are jet-black, some are white, some are brown; big cats of every description and colour are seen by some unsuspecting person while on his or her daily business. 'Big Cats in Britain' are the largest and most active research group in the British Isles and Ireland. This book contains a run-down of every known big cat sighting in the UK during 2006, together with essays by various luminaries of the British big cat research community.

CAT FLAPS! Northern Mystery Cats
Andy Roberts - ISBN 978-1-905723-11-9

£6.99

Of all Britain's mystery beasts, the alien big cats are the most renowned. In recent years the notoriety of these uncatchable, out-of-place predators have eclipsed even the Loch Ness Monster. They slink from the shadows to terrorise a community, and then, as often as not, vanish like ghosts. But now film, photographs, livestock kills, and paw prints show that we can no longer deny the existence of these once-legendary beasts. Here then is a case-study, a true lost classic of Fortean research by one of the country's most respected researchers.

CENTRE FOR FORTEAN ZOOLOGY 2007 YEARBOOK
Edited by Jonathan Downes and Richard Freeman
ISBN 978-1-905723-14-0

£12.50

The Centre For Fortean Zoology Yearbook is a collection of papers and essays too long and detailed for publication in the CFZ Journal *Animals & Men.* With contributions from both well-known researchers, and relative newcomers to the field, the Yearbook provides a forum where new theories can be expounded, and work on little-known cryptids discussed.

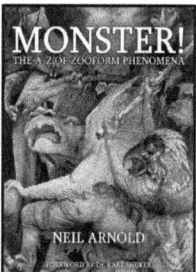

MONSTER! THE A-Z OF ZOOFORM PHENOMENA
Neil Arnold - ISBN 978-1-905723-10-2

£14.99

Zooform Phenomena are the most elusive, and least understood, mystery `animals`. Indeed, they are not animals at all, and are not even animate in the accepted terms of the word. Author and researcher Neil Arnold is to be commended for a groundbreaking piece of work, and has provided the world's first alphabetical listing of zooforms from around the world.

CFZ PRESS, MYRTLE COTTAGE,
WOOLFARDISWORTHY BIDEFORD,
NORTH DEVON, EX39 5QR
www.cfz.org.uk

Other books available from
CFZ PRESS

BIG CATS LOOSE IN BRITAIN
Marcus Matthews - ISBN 978-1-905723-12-6

£14.99

Big Cats: Loose in Britain, looks at the body of anecdotal evidence for such creatures: sightings, livestock kills, paw-prints and photographs, and seeks to determine underlying commonalities and threads of evidence. These two strands are repeatedly woven together into a highly readable, yet scientifically compelling, overview of the big cat phenomenon in Britain.

DARK DORSET
TALES OF MYSTERY, WONDER AND TERROR
Robert. J. Newland and Mark. J. North
ISBN 978-1-905723-15-6

£12.50

This extensively illustrated compendium has over 400 tales and references, making this book by far one of the best in its field. Dark Dorset has been thoroughly researched, and includes many new entries and up to date information never before published. The title of the book speaks for itself, and is indeed not for the faint hearted or those easily shocked.

MAN-MONKEY - IN SEARCH OF THE BRITISH BIGFOOT
Nick Redfern - ISBN 978-1-905723-16-4

£9.99

In her 1883 book, *Shropshire Folklore*, Charlotte S. Burne wrote: *'Just before he reached the canal bridge, a strange black creature with great white eyes sprang out of the plantation by the roadside and alighted on his horse's back'*. The creature duly became known as the `Man-Monkey`.

Between 1986 and early 2001, Nick Redfern delved deeply into the mystery of the strange creature of that dark stretch of canal. Now, published for the very first time, are Nick's original interview notes, his files and discoveries; as well as his theories pertaining to what lies at the heart of this diabolical legend.

EXTRAORDINARY ANIMALS REVISITED
Dr Karl Shuker - ISBN 978-1905723171

£14.99

This delightful book is the long-awaited, greatly-expanded new edition of one of Dr Karl Shuker's much-loved early volumes, *Extraordinary Animals Worldwide*. It is a fascinating celebration of what used to be called romantic natural history, examining a dazzling diversity of animal anomalies, creatures of cryptozoology, and all manner of other thought-provoking zoological revelations and continuing controversies down through the ages of wildlife discovery.

**CFZ PRESS, MYRTLE COTTAGE,
WOOLFARDISWORTHY BIDEFORD,
NORTH DEVON, EX39 5QR
w w w . c f z . o r g . u k**

Other books available from
CFZ PRESS

DARK DORSET CALENDAR CUSTOMS
Robert J Newland - ISBN 978-1-905723-18-8

£12.50

Much of the intrinsic charm of Dorset folklore is owed to the importance of folk customs. Today only a small amount of these curious and occasionally eccentric customs have survived, while those that still continue have, for many of us, lost their original significance. Why do we eat pancakes on Shrove Tuesday? Why do children dance around the maypole on May Day? Why do we carve pumpkin lanterns at Hallowe'en? All the answers are here! Robert has made an in-depth study of the Dorset country calendar identifying the major feast-days, holidays and celebrations when traditionally such folk customs are practiced.

CENTRE FOR FORTEAN ZOOLOGY 2004 YEARBOOK
Edited by Jonathan Downes and Richard Freeman
ISBN 978-1-905723-14-0

£12.50

The Centre For Fortean Zoology Yearbook is a collection of papers and essays too long and detailed for publication in the CFZ Journal *Animals & Men*. With contributions from both well-known researchers, and relative newcomers to the field, the Yearbook provides a forum where new theories can be expounded, and work on little-known cryptids discussed.

CENTRE FOR FORTEAN ZOOLOGY 2008 YEARBOOK
Edited by Jonathan Downes and Corinna Downes
ISBN 978-1-905723-19-5

£12.50

The Centre For Fortean Zoology Yearbook is a collection of papers and essays too long and detailed for publication in the CFZ Journal *Animals & Men*. With contributions from both well-known researchers, and relative newcomers to the field, the Yearbook provides a forum where new theories can be expounded, and work on little-known cryptids discussed.

ETHNA'S JOURNAL
Corinna Newton Downes
ISBN 978-1-905723-21-8

£9.99

Ethna's Journal tells the story of a few months in an alternate Dark Ages, seen through the eyes of Ethna, daughter of Lord Edric. She is an unsophisticated girl from the fortress town of Cragnuth, somewhere in the north of England, who reluctantly gets embroiled in a web of treachery, sorcery and bloody war...

CFZ PRESS, MYRTLE COTTAGE,
WOOLFARDISWORTHY BIDEFORD,
NORTH DEVON, EX39 5QR
www.cfz.org.uk

Other books available from
CFZ PRESS

ANIMALS & MEN - Issues 11 - 15 - The Call of the Wild
Jonathan Downes (Ed) - ISBN 978-1-905723-07-2

£12.50

Since 1994 we have been publishing the world's only dedicated cryptozoology magazine, *Animals & Men*. This volume contains fascimile reprints of issues 11 to 15 and includes articles covering out of place walruses, feathered dinosaurs, possible North American ground sloth survival, the theory of initial bipedalism, mystery whales, mitten crabs in Britain, Barbary lions, out of place animals in Germany, mystery pangolins, the barking beast of Bath, Yorkshire ABCs, Molly the singing oyster, singing mice, the dragons of Yorkshire, singing mice, the bigfoot murders, waspman, British beavers, the migo, Nessie, the weird warbling whatsit of the westcountry, the quagga project and much more...

IN THE WAKE OF BERNARD HEUVELMANS
Michael A Woodley - ISBN 978-1-905723-20-1

£9.99

Everyone is familiar with the nautical maps from the middle ages that were liberally festooned with images of exotic and monstrous animals, but the truth of the matter is that the *idea* of the sea monster is probably as old as humankind itself.

For two hundred years, scientists have been producing speculative classifications of sea serpents, attempting to place them within a zoological framework. This book looks at these successive classification models, and using a new formula produces a sea serpent classification for the 21st Century.

**CFZ PRESS, MYRTLE COTTAGE,
WOOLFARDISWORTHY BIDEFORD,
NORTH DEVON, EX39 5QR
www.cfz.org.uk**

www.ingramcontent.com/pod-product-compliance
Lightning Source LLC
Chambersburg PA
CBHW062156080426
42734CB00010B/1716